AF838497

Padang, West Sumatra, Indonesia Earthquake of 2009

Lifeline Performance

PREPARED BY
Earthquake Investigation Committee of the Technical Council of Lifeline Earthquake Engineering

EDITED BY
Alex K. Tang, P.E., P.Eng., C.Eng., F.ASCE

SPONSORED BY
Technical Council on Lifeline Earthquake Engineering
Monograph No. 37

ASCE

Published by the American Society of Civil Engineers

Library of Congress Cataloging-in-Publication Data

Padang, West Sumatra, Indonesia earthquake of 2009 lifeline performance / prepared by Earthquake Investigation Committee of the Technical Council of Lifeline Earthquake Engineering ; edited by Alex K. Tang ; sponsored by Technical Council on Lifeline Earthquake Engineering.
pages cm. -- (Monograph / Technical Council on Lifeline Earthquake Engineering ; no. 37)
Includes bibliographical references and index.
ISBN 978-0-7844-1293-0 (paper) -- ISBN 978-0-7844-7788-5 (pdf)
1. Lifeline earthquake engineering--Indonesia--Padang--Evaluation. 2. Earthquakes--Indonesia--Padang. 3. Emergency management--Indonesia--Padang. 4. Earthquake hazard analysis--Indonesia--Padang. I. Tang, Alex. II. American Society of Civil Engineers. Earthquake Investigations Committee. III. American Society of Civil Engineers. Technical Council on Lifeline Earthquake Engineering.
TA654.6.P25 2013
624.1'7620959813--dc23
2013022817

Published by American Society of Civil Engineers
1801 Alexander Bell Drive
Reston, Virginia 20191
www.asce.org/pubs

ISBN 978-0-7844-1293-0 (print)
ISBN 978-0-7844-7788-5 (PDF)
Manufactured in the United States of America.

TCLEE Monograph Series

Publications in print may be purchased from ASCE via telephone at 1-800-548-ASCE (2723) or online at www.asce.org/bookstore. The TCLEE Web site is www.asce.org/tactclee.

No. 1 Recent Lifeline Seismic Risk Studies, Kiremidjian, Anne S., Editor, 1990.

No. 2 Seismic Lost Estimates for a Hypothetical Water System, A Demonstration Project, Taylor, Craig E., Editor, August 1991.

No. 3 Guide to Post-Earthquake Investigations of Lifelines, Schiff, Anshel J., Editor, August 1991.

No. 4 Lifeline Earthquake Engineering, Proceedings of the 3rd U.S. Conference on Lifeline Earthquake Engineering, August 22-23, 1991, Los Angeles, CA, Cassaro, Michael, Editor, August 1991.

No. 5 Lifeline Earthquake Engineering in the Central and Eastern United States, Ballantyne, Donald, Editor, September 1992.

No. 6 Lifeline Earthquake Engineering, Proceeding of the 4th U.S. Conference on Lifeline Earthquake Engineering, August 10-12, 1995, San Francisco, CA, O'Rourke, Michael J., Editor, August 1995.

No. 7 Critical Issues and State of the Art on Lifeline Earthquake Engineering, Schiff, Anshel J. and Buckle, Ian, Editors, October 1995.

No. 8 Northridge Earthquake: Lifeline Performance and Post-Earthquake Response, Schiff, Anshel J., Editor, August 1995.

No. 9 Seismic Design for Natural Gas Distributors, McDonough, Peter W., August 1995.

No. 10 Methods of Achieving Improved Seismic Performance of Communications Systems, Tang, Alex, and Schiff, Anshel J., Editors, September 1996.

No. 11 Guide to Post-Earthquake Investigation of Lifelines, Schiff, Anshel J., Editor, July 1997.

No. 12 Seismic Guidelines for Ports, Werner, Stuart D., Editor, March 1998.

No. 13 Overcoming Barriers: Lifeline Seismic Improvement Programs, Taylor, Craig E., Mittler, Elliott, and Lund, Le Val, September 1998.

No. 14 Hyogo-Ken Nambu Earthquake of January 17, 1995—Lifeline Performance, Schiff, Anshel J. Editor, 1998.

No. 15 Guidelines for the Seismic Evaluation and Upgrade of Water Transmission Facilities, Eidinger, John M. and Avila, Ernesto A., Editors, January 1999.

No. 16 Optimizing Post-Earthquake Lifeline System Reliability (Proceedings of the 5th U.S. Conference on Lifeline Earthquake Engineering, Seattle, Washington, August 12-14, 1999.), Elliott, William M., and McDonough, Peter W., Editors, August 1999.

No. 17 Ismit (Kocaeli), Turkey Earthquake of August 16, 1999, Including Duzce Earthquake of November 12, 1999—Lifeline Performance, Tang, Alex K., Editor, September 2000.

No. 18 Chi-Chi, Taiwan, Earthquake of September 21, 1999—Lifeline Performance, Schiff, Anshel J., and Tang, Alex K., Editors, October 2000.

No. 19 Gujarat (Kutch) India, M7.7 Earthquake of January 26, 2001 and NAPA M5.2 Earthquake of September 3, 2000, Eidinger, John M., Editor, June 2001.

No. 20 The Nisqually, Washington, Earthquake of February 2001—Lifeline Performance, McDonough, Peter W., Editor, February 2002.

No. 21 Acceptable Risk Process—Lifelines and Natural Hazards, Taylor, Craig E., and VanMarcke, Erik H., Editors, March 2002.

No. 22 Seismic Screening Checklists for Water and Wastewater Facilities, Heubach, William F., Editor, September 2002.

No. 23 Atico, Peru Mw 8.4 Earthquake of June 23, 2001, Curtis L. Edwards, Editor, October 2002.

No. 24 Lifeline Performance of El Salvador Earthquakes of January 13 and February 13, 2001, Le Val Lund, Editor and Carl Sepponen, Editor, September 2002.

No. 25 Advancing Mitigation Technologies and Disaster Response for Lifeline System: Proceedings of the Sixth U.S. Conference and Workshop on Lifeline Earthquake Engineering, Beavers, J.E., Editor, August 2003.

No. 26 Fire Following Earthquake, Scawthorn, Charles, Eidinger, John M., and Schiff, Anshel J., Editors, 2005.

No. 27 Zemmouri, Algeria, Mw 6.8 Earthquake of May 31, 2003, Edwards, Curtis L., Editor, 2004.

No. 28 San Simeon Earthquake of December 22, 2003 and Denali, Alaska, Earthquake of November 3, 2002, Yashinsky, Mark, Editor, Lund, Le Val, Co Editor, 2004.

No. 29 Hurricane Katrina: Performance of Transportation Systems, DesRoches, Reginald, Editor, 2006.

No. 30 Sumatra-Andaman Islands Earthquake and Tsunami of December 26, 2004 Lifeline Performance, Strand, Carl and Masek, John, Editors, 2007.

No. 31 Kashiwazaki, Japan Earthquake of July 16, 2007 Lifeline Performance, Tang, Alex K.K. and Schiff, Anshel J., Editors, 2007.

No. 32 Pisco, Peru Earthquake of August 15, 2007 Lifeline Performance, Tang, Alex K.K. and Johansson, Jorgen, Editors, 2007.

No. 33 Pacific Northwest Storms of December 1-4, 2007: Lifeline Performance, edited by Teresa Elliott, P.E., and Alex K. Tang, P.E.

No. 34 Seismic Resilience of Natural Gas Systems: Improving Performance, edited by Peter W. McDonough, P.E.

No. 35 Haiti M_w 7.0 Earthquake of January 12, 2010: Lifeline Performance, edited by Curtis L. Edwards, P.E.

No. 36 Chile Earthquake of 2010: Performance of Lifelines, edited by Alex K. Tang and John Eidinger.

No. 37 Padang, West Sumatra, Indonesia, Earthquake of 2009: Lifeline Performance, edited by Alex K. Tang

Other TCLEE Publications

Duke, C. Martin, Editor, The Current State of Knowledge of Lifeline Earthquake Engineering, Proceedings TCLEE Specialty Conference, August 30-31, 1977, Los Angeles, CA, (Later designated as the 1st U.S. Conference on Lifeline Earthquake Engineering), August, 1977.

Dowd, Munson, Editor, Annotated Bibliography on Lifeline Earthquake Engineering, 1980.

Smith, D. J, Jr., Editor, Lifeline Earthquake Engineering, The Current State of Knowledge 1981, Proceedings of the Second TCLEE Specialty Conference, August 20-21, 1981, Oakland, CA, (Later designated as the 2nd U.S. Conference on Lifeline Earthquake Engineering), August 1981.

Hall, William J., Advisory Notes on Lifeline Earthquake Engineering, 1983.

Nyman, Douglas, NSF Principal Investigator, Guidelines for the Seismic Design of Oil and Gas Pipelines Systems, TCLEE Committee on Gas and Liquid Fuels, 1984.

Cooper, James, Editor, Lifeline Earthquake Engineering Performance, Design and Construction, 1984.

Cassaro, Michael and Martinez-Romero, E., Editors, The Mexico Earthquake, 1985, Factors Involved and Lessons Learned, 1986.

Eguchi, Ronald and Crouse, C. B., Lifeline Seismic Risk Analysis - Case Studies, 1986.

Wang, Leon R. L. and Whitman, Robert, Seismic Evaluation of Lifeline Systems-Case Studies, 1986.

Cassaro, Michael and Cooper, James, Editors, Seismic Design and Construction of Complex Civil Engineering Systems, 1988.

Werner, Stuart D. and Dickenson, Stephen E., Editors, Hyogo-Ken Nambu (Kobe) Earthquake of January 17, 1995: A Post-Earthquake Reconnaissance of Port Facilities, TCLEE Ports Committee, 1996.

ASCE Manual

Schiff, Anshel J., Editor, Guide to Improved Earthquake Performance of Electric Power Systems, ASCE Manual 96.

TCLEE Earthquake Investigation Reports

TCLEE has also prepared numerous earthquake reports that have appeared in other publications. References to these reports and 10 short reports associated with TCLEE monographs can be viewed on the ASCE/TCLEE Web site address given below. The 10 short reports are each about 5 to 15 pages long and contain a summary of main observations and some pictures. They can be downloaded at *www.asce.org/tactclee*.

CONTENTS

PREFACE

The Earthquake Investigation Committee of the Technical Council of Lifeline Earthquake Engineering (TCLEE), American Society of Civil Engineers (ASCE), was established to initiate, organize, train for, coordinate, and evaluate the performance of lifelines following earthquakes. Members of the committee are employees of lifeline industries, consulting engineers, and academics from the United States and Canada. Committee members provide services on a voluntary basis. For some earthquake investigations, participants' companies do not require an individual to take vacation time for the investigation and may provide some support for expenses. ASCE also provides support to reimburse expenses. In addition to the time associated with the reconnaissance trip, the substantial effort by each individual to prepare a short report for the TCLEE Web page and the full report for the monograph series is all done on a voluntary basis. The cost of this effort is substantially more than the support provided by ASCE.

Individuals participating in the investigation need not be members of the committee or ASCE, but they are expected to follow the committee's earthquake investigation practices as described in the ASCE publication, TCLEE Monograph 11, "Guide to Post-Earthquake Investigation of Lifelines." Members of the investigation team coordinate with other groups and may participate in groups organized by other organizations. They gather both good and poor performance data from domestic and foreign earthquakes to provide information for practitioners to improve the performance of the lifeline systems. Foreign earthquakes that have been investigated include the 1985 Chile, 1988 Soviet Armenia, 1990 Philippines, 1991 Costa Rica, 1992 Turkey, 1994 Kobe (Japan), 1999 Kocaeli (Turkey), 1999 Chi-Chi (Taiwan), 2001 Gujarat (India), 2002 Atico (Peru), 2004 Zemmouri (Algeria), 2004 Sumatra, 2007 Kashiwazaki (Japan), 2008 Pisco (Peru) earthquakes, 2008 Wenchuan (China) earthquake, and 2009 L'Aquila (Italy) earthquake. Domestic earthquakes that have been investigated include the 1989 Loma Prieta, 1992 Landers, 1994 Northridge, 2000 Napa, 2001 Nisqually, 2002 Denali, 2003 Paso Robles, and 2008 Alum Rock earthquake.

The Kobe earthquake report, Monograph 14, was the first foreign earthquake investigation report published by ASCE as a TCLEE monograph. The first domestic earthquake investigation report published by ASCE as a TCLEE monograph, number 8, was for the Northridge earthquake. Prior to this time, TCLEE prepared a lifeline report that was published by the Earthquake Engineering Research Institute (EERI). The Earthquake Investigation Committee continues to cooperate with EERI to provide an abbreviated version of lifeline performance in Earthquake Spectra (EERI publication). TCLEE publishes brief preliminary reports on the ASCE/TCLEE Web page.

Alex Tang, PE, F. ASCE

CONTRIBUTORS

Name	Affiliation	E-mail
Sudi Shoja	DSS Services	Sudi@dsservices.org
Teuku Alvisyahrin	Tsunami and Disaster Mitigation Research Center, Banda Aceh, Indonesia (TDMRC), Syiah Kuala University	alvis@tdmrc.org
Alex Tang	L&T Consultant	alexktang@mac.com

ACKNOWLEDGMENTS

Most of the contributing authors of this report are members of the Earthquake Investigation Committee (EIC) of Technical Council on Lifeline Earthquake Engineering (TCLEE). Sudi Shoja and Alex Tang were funded by ASCE. ASCE support was received at the direction of Pat Natale, ASCE executive director, with support from John Durrant, John Segna, and Tenzing Barshee. Unanimous decision to approve the trip was granted by the ASCE Executive Committee. The investigation team is very grateful to Professor H. S. Oey of the University of Texas, El Paso, and Dr. Talib Puspokusumo, who provided us with contacts to visit lifeline facilities and collect relevant performance information. We also would like to thank the mayor of the city of Padang, Dr. H. Fauzi Bahar, and his assistant, Dr. Didi Aryadi, who provided us with valuable support to meet with the city department heads. We are also indebted to Mr. Teddy H. Yudha, Mr. Eri F. Arsya, Mr. Roni Handawi, and Mr. Frans Jaya Nio, who established all local logistics for us including locating a safe hotel and car rental. Mr. Roni Handawi also set up meetings with telecom services providers in Padang and visits to the equipment buildings. Mr. Eri F. Arsya arranged meetings and contacts with the Public Works Department and Andalas University. Dr. Talib Puspokusumo, Mr. Roni Handawi, and Mr. Eri F. Arsya, all from Jakarta, stayed with us in Padang for 2 days.

Local officials, such as the fire department and water supply department, should be commended for the time they took from their busy restoration/relief efforts to provide us with data and information contained in this report.

As is not uncommon in post-earthquake reconnaissance, incomplete information in the weeks and months after the event can lead to omissions and misunderstandings. We apologize if the findings in this report are incomplete. It may take months to years of post-earthquake evaluations before a comprehensive understanding of lifeline impacts is available. ASCE/TCLEE will continue to update the lifeline performance information when more information is collected.

The findings and photos presented in this report reflect the collected input from many people and sources. The ASCE TCLEE team of Sudi Shoja, and Alex Tang collected site data from October 26 to October 30, 2009.

I would like to dedicate this monograph to Dr. Talib Puspokusumo, who had a heart attack and died in early 2012, for his effort and belief in loss reduction in disasters.

Alex K. Tang, P.E., P. Eng., C. Eng., F. ASCE

The Mayor of Padang met with us at his office. Left to right, Sudi Shoja (ASCE/TCLEE), Dr. H. Fauzi Bahar (Mayor of Padang), Dr. Talib Puspokusumo (Former Consulate General of Indonesia in the United States), and Alex Tang (ASCE/TCLEE).

ASCE/TCLEE team leader Alex Tang presents an investigation vest to the Mayor of Padang.

We were taken to the only restaurant open for business along the beach the first night in Padang after completing all scheduled meetings for the day. Frans Jaya Nio (right) who helped us with local logistics joined us after we started.

Left to right, Dr. Nudhi Erwanto, head of Fire Department, Alex Tang (ASCE/TCLEE), and Basil ST, second in Command of the Fire Department of Padang, after the meeting with the Pandang Fire Department heads.

Left to right, Diky Juliand (TDMRC), Sudi Shoja (ASCE/TCLEE), Henky Mayaguezz (Emergency Response), Dedi Hanidal (Manager of Emergency Response), Alex Tang (ASCE/TCLEE), Teuku Alvisyahrin (TDMRC, Head of Professional Services Division), Wan Akmal (TDMRC) at the Emergency Response control center.

Left to right, Sudi Shoja (ASCE/TCLEE), Alex Tang (ASCE/TCLEE), and Syahril, ST (General Manager Telkom) after surveying one of the two damaged Telkom office building. These are not equipment offices.

From left to right, Roni Handawi (PT. Sanex Intim Infokom), Thomas (Manager Network Operations, Telkomsel), and Alex Tang (ASCE/TCLEE) after the meeting at Telkomsel office in Padang. Heriyanto spent more than 2.5 hours providing Tang with photos of damage and wireless system performance information.

Left to right, Dr. Talib Puspokusumo, Sudi Shoja (ASCE/TCLEE), Eri F. Arsya of RPX, Ir. H. Azhar Letif Chief of Padang Water Department, and Alex Tang (ASCE/TCLEE).

Left to right, Dr. Talib Puspokusumo, Ir. H. Azhar Letif Chief of Padang Water Department, Sudi Shoja (ASCE/TCLEE), and Harry Satria, Operations Manager Padang Water Department.

Left to right, Diky Juliand (TDMRC), Wan Akmal (TDMRC), Sudi Shoja (ASCE/TCLEE), Alex Tang (ASCE/TCLEE), Teuku Alvisyahrin (TDMRC, Head of Professional Services Division), and Purwandy Hasibuan (TDMRC) after a meeting with the engineers of the building inspection group leader at Andalas University, Padang.

Chapter 1
INTRODUCTION

On September 30, 2009 at 5:16 p.m. local time, there was a magnitude (M_w) 7.6 earthquake off the west coast of the island of Sumatra in Indonesia, about 60 km NW of Padang, a city in West Sumatra Province (Fig. 1.1). An aftershock of magnitude 6.2 occurred within 22 minutes of the main event; numerous aftershocks within the first 4 weeks had magnitudes ranging from 3 to 5. The aftershock caused panic among the residents of Padang, as many structures damaged by the main shock collapsed due to the aftershocks. The social economic impact of the earthquake was significant with a loss of 1,100 lives and more than 2,000 injuries. About 50 percent of all residential buildings collapsed or were deemed unsafe; many churches, schools, and hospitals were also damaged. Almost all government-owned buildings sustained various degrees of damage—becoming unsafe for occupancy or collapsing. The districts affected by the earthquake are shown in Figure 1.2. The estimated total losses were about $2.06 billion (U.S.) including lifeline infrastructure losses. A very small-scale tsunami occurred at the coast of Padang. No damage was reported; however, surfers were reported missing.

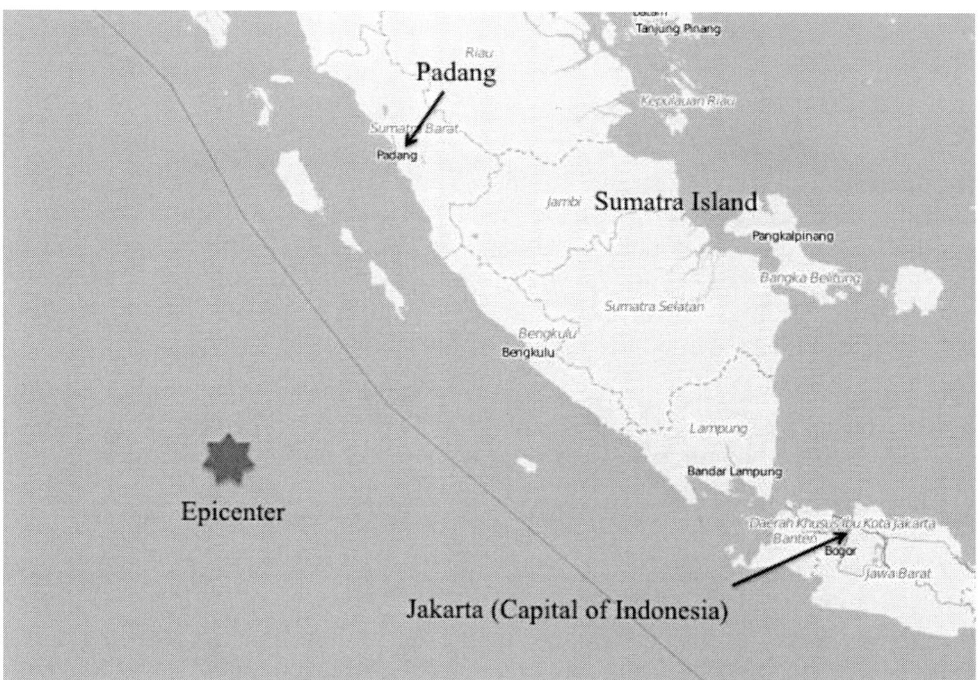

Figure 1.1. Areas affected by the September 30 earthquake
(Source: OpenStreetMap, available under the Open Database License)

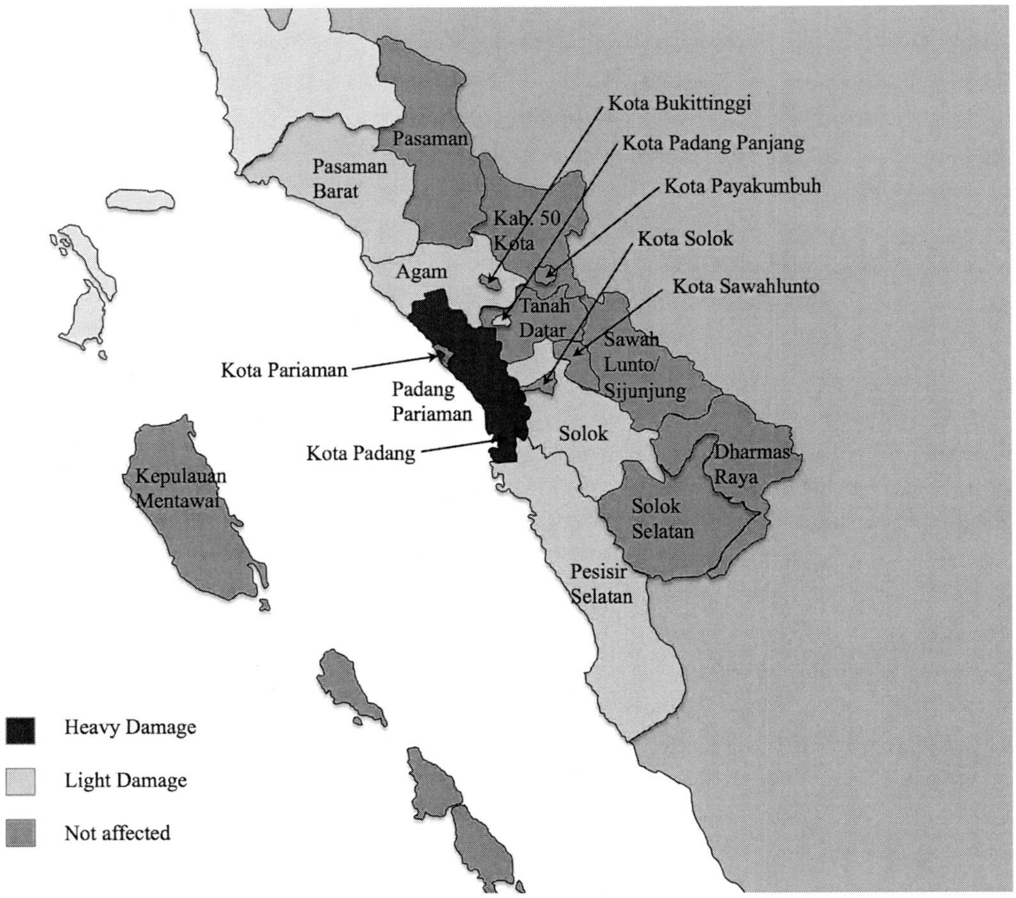

Figure 1.2. Districts affected by the September 30 earthquake

The earthquake occurred between the Sumatra Fault along the west coast of Sumatra and the subduction plate boundary west of the islands off the coast of Sumatra (Fig. 1.2). The fault responsible for this earthquake is situated between the Sumatra Fault and the plate boundary. It is called the Mentawai Fault and was discovered by Professor Badrul Mustafa Kemal of Andalas University, Padang. We met Professor Kemal on the second day of our visit, and he explained the details of his findings to us.

This plate boundary and the fault system of Sumatra are the source of numerous significant earthquakes in this region since 1900 (Fig. 1.3). The strong shaking coupled with strong aftershocks resulted in destruction of many adobe, brick masonry, brick in-filled concrete frame buildings, and steel buildings lacking joint details as well as severe damage to lifelines. The recorded maximum peak ground acceleration along the coast of Padang was 0.4 g.

Many hotels in the city sustained various degrees of damage. More than 60 percent were closed due to sustained structural damage or partial collapsed. A few of these hotels are starred hotels. One hospital building and one school collapsed. Many schools were unsafe and were closed.

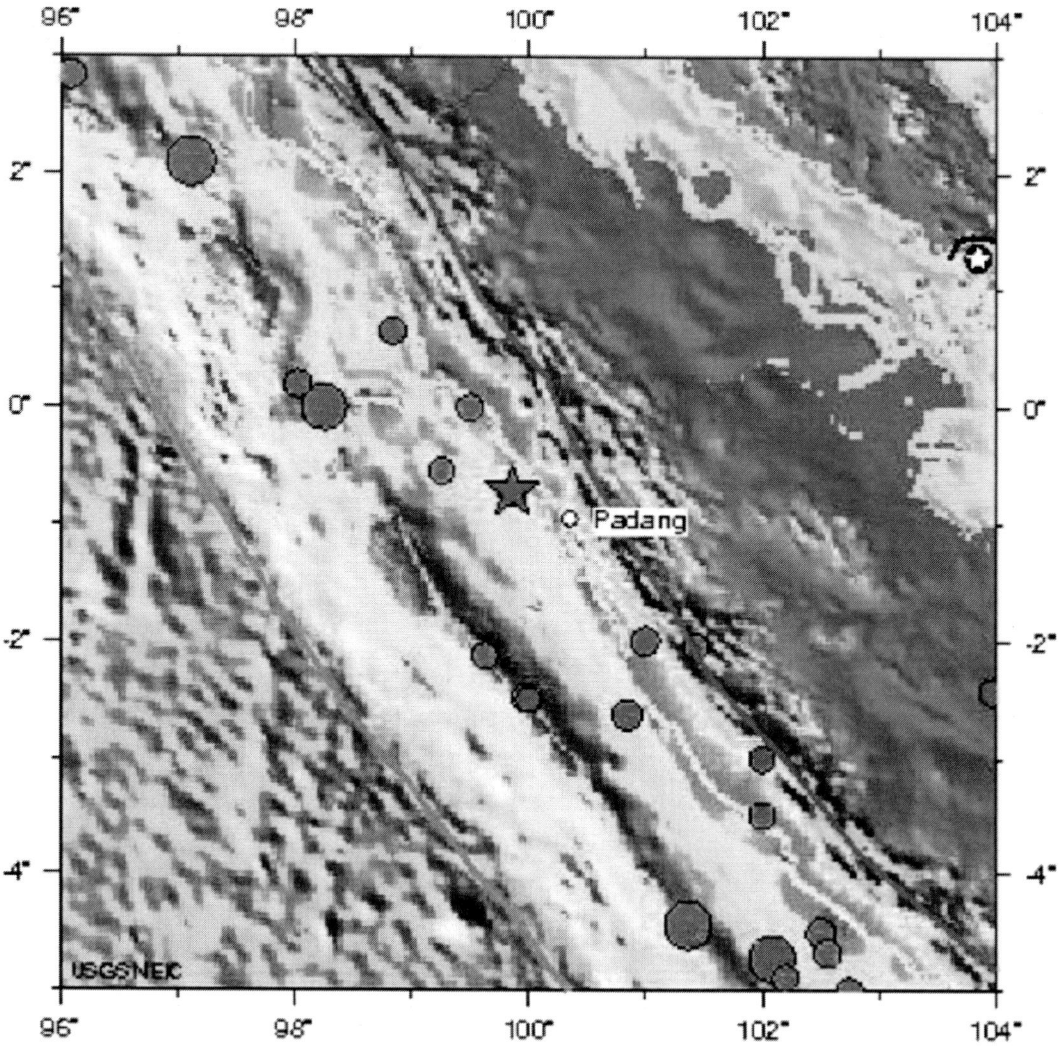

Figure 1.3. Earthquakes of magnitude 7 and higher since 1900 in the region. Note that most of the events occurred between the fault system in Sumatra and the plate boundary west of the island. (Courtesy of USGS)

The power disruption in the affected areas ranged from a few hours to 3 days. Many cell sites without battery reserve power went down. As a result, many sections of the damaged area were without cellular phone services for more than 8 hours.

Communications systems, particularly cellular networks, performed poorly the first few days after the earthquake. Cellular phones are more commonly used in the region. The damage to batteries in the central office along with power outages resulted in landline service disruption. There was damage to the central offices' telecommunication equipment circuit packs. Landline distribution system damage was limited to aerial cables due to damaged poles. The loss of telecommunications immediately after the earthquake rendered emergency response and rescue efforts more difficult.

The water system sustained heavy damage due to collection pipe breaks at the water source and pipe breaks in the distribution network. The water system was operating at 50 percent of its capacity 5 weeks after the earthquake. Water tankers and temporary water tanks were used in the city of Padang to distribute potable water. In areas outside the city, small water purification units were used to purify water from streams to provide water for the local residents. There is no wastewater treatment plant in the area; septic tanks are used.

The roads and bridges sustained light damage, surface cracks, and bridge abutment settlement. Most of the repairs were done within a couple of weeks of the earthquake.

Both hospitals and schools sustained significant damage to the buildings and loss of lifeline services. One building in a government hospital complex collapsed, while the non-structural elements inside the main building were severely damaged.

The heavily damaged areas of Padang caught fire following earthquake. Luckily, the rain helped to prevent the fire from spreading. It took fire fighters 3 days to subdue one major fire. The fire department building roof collapsed, damaging one fire truck that was out of service. There were not enough fire trucks to handle all the fires.

Indonesian military and international rescue teams performed well in both emergency response and rescue. Aid such as food and medical supplies from many countries and non-governmental organizations (NGOs) helped to relieve hardships due to the earthquake.

Power Systems

The electric power distribution network in earthquake-impacted areas sustained extensive damage. The modes of failure were toppled utility poles, pole mounted transformers falling or sustaining internal damage, and drop wires that were disconnected due to partial building failure or utility pole movement. The latter was the main reason for prolonged electric power outage in the city.

The power outage caused the cellular network malfunction, which hindered emergency response and rescue during the hours immediately after the earthquake.

Communications

Extensive damage to both the landline network and the cellular network was observed. Switching center building damage, battery bank collapse, equipment circuit board assemblies damage/malfunction, and severed landlines were the causes of landline communication interruption. Temporary fixes restored service to majority of the customer base in Padang within 2 days of the earthquake.

Mobile cell sites were deployed where the original cell sites (base transceiver stations [BTS]) were collapsed or antennae towers were damaged or toppled. The cellular connection was sporadic during our investigation 4 weeks after the earthquake. Mobile power generators were also deployed to undamaged cell sites to power the equipment during the power outage. Most BTSs has less than 3 hours of battery reserve power, while in some cases the batteries were damaged.

Telkom, the major telecommunication provided in Indonesia, sent equipment and technical staff from the Jakarta head office to Padang to support network recovery.

Water Systems

The water intake systems in two out of five water treatment plants serving Padang had broken pipes. Equipment inside one plant also suffered minor damage.

The distribution system sustained severe setbacks with more than 1,900 pipe breaks. The water supply was only about 50 percent of its original capacity four weeks after the earthquake. Water tanks and water trucks, small water purification units, and seawater desalination unit were deployed to provide potable water to the community.

Due to turbidity after the earthquake, most wells were closed.

The Water Department head office had structural damage and was deemed unsafe; all office staff members were working outside the building.

Highways and Bridges

The highways, roads, and bridges within the earthquake area had minor setbacks, such as surface cracks, permanent ground deformation, and landslides. Most surface cracks were repaired within a couple of weeks of the earthquake. Temporary fixes were observed at five damaged bridge abutments that were reported by the Transportation Department. No leveling was done at either end of the abutments; automobiles had to slow down as if crossing a speed bump. Bridge railings suffered minor damage but were not repaired.

A steel water transmission pipe (about 20 cm in diameter) colocated with a damaged bridge was broken; it was welded within 2 weeks of the earthquake.

Sea Ports and Airports

There was no damage to the seaport of Teluk Buyur, while Padang's international airport was closed for about 12 hours because a part of the terminal building roof was damaged and debris scattered inside the terminal. The airport reopened the morning of October 1, 2009.

Gas and Tanks

A helicopter survey showed that there was no damage to the tank farm in the seaport of Teluk Buyur (Fig. 1.4).

Hospitals and Schools

Two hospitals had significant damage to their buildings, and a building collapsed at a third. Tents were set up outside the hospital to handle the patients and minor injuries. One main hospital building had extensive ceiling fixtures down, file cabinets with files all over the floor, and damaged light fixtures.

Figure 1.4. Aerial views of Teluk Buyur port, the tank farms, and the silos showed no signs of damage. This photo was taken facing north.
(Courtesy of Pengiran Yusuf Ederis)

One college building collapsed, and most school buildings in Padang sustained some degree of damage. After surveying the schools, the mayor of Padang proposed funding to rebuild all schools and to turn them into safe houses in case of future tsunami.

Emergency Response, Recovery, and Social Impact

The lifeline companies executed their own emergency response plans to survey and restore damage to maintain some service. Most took 3 to 5 hours to assess the extent of damage before establishing a plan to restore service. By that time, nightfall was approaching, which increased the difficulties in restoration. Traffic lights at intersections were out due to loss of power, causing traffic congestion and making it more difficult to get to the sites where repair was required.

The military was called in to support emergency response to collapsed buildings. International support arrived within 2 days. A few countries sent emergency response personnel with equipment to help Indonesian military personnel. Australia sent desalinating equipment with power generators and set them up along the coast to provide potable water to the community.

Within 3 to 5 days, most of the lifeline services had been restored to an acceptable level. It will take a few months to restore normal service levels.

Many family businesses in the city suffered extensively in this earthquake. Many operate in two-story buildings with the business on the ground floor and family living quarters on the second floor. As a result, they lost both their business operations and their homes—their income and their residences. One print shop was totally destroyed. A second collapsed, and all the printing equipment on the ground floor was damaged. Fortunately, the owner and his wife ran outside before the second floor collapsed. Their children were at school, and the school building had only minor damage.

Many restaurants along the coast were damaged, and some collapsed. The coast along Padang is a tourist area, and it was very quiet while we were there 4 weeks after the earthquake.

The tourist industry in Padang will lose millions of dollars (U.S.) due to extensive damage to the star hotels, beachfront properties, and restaurants. It will be difficult to estimate the total economic loss until Padang is fully recovered. The citizens of Padang, however, are motivated to rebuild the city to its original prosperity.

Organization

This report discusses the following lifeline systems: electric power, communication, water, highways, and emergency response. Lifeline systems within hospitals are also discussed. Most lifeline chapters start with an overview of the system performance, followed by sections that describe the system, its damage, emergency response, and recovery. Each chapter finishes with a conclusions and recommendations.

Chapter 2
GEOTECH AND GEOLOGY

On September 30, 2009, another massive earthquake hit Padang, Sumatra, Indonesia at 10:16:09 UTC, or 05:16:09 p.m. local time at the epicenter. The earthquake had a moment magnitude of 7.6 on the Richter scale and was located at 0.84°S 99.65°E, according to the Indonesian Agency for Meteorology, Climatology and Geophysics/BMKG. It caused severe damage in Padang and Pariaman, as the epicenter was only about 57 km southwest of these cities.

2.1 Brief Overview of Tectonic System

This calamity should not be surprising if we consider the tectonic system involved. Indonesia is located at the confluence of three active earth tectonic plate boundaries: the Eurasian Plate, the Indo-Australian Plate, and the Pacific Plate (Fig. 2.1). This condition causes most regions in Indonesia to be highly prone to tectonic earthquakes. In general, the most vulnerable areas include Sumatra, Java, East Nusa Tenggara, Maluku, Papua, and Sulawesi.

McCaffrey (1996) stated that Indonesia is a place where active tectonic plates converge. The Indo-Australian plate, for example, strikes straight across Java Island (frontal strike), while the Sumatra Island has side (oblique) strike characteristics. This phenomenon triggers active faults along Sumatra Island.

With respect to the area of Sumatra, the devastating earthquakes that have occurred as well as the ongoing tectonic processes throughout the region have resulted in a potential for future earthquakes. In particular, there is a great potential for an inland earthquake due to faults, which directly or indirectly have accommodated the energy from seismic activity in the subduction zone as well as from continuous plate movements.

The latest and biggest seismic activity has occurred on Nias and Simeulue islands, causing major earthquakes in Sumatra. Great Sumatran Fault (GSF) splits Sumatra island dominantly moving towards strike-slip to the right (Sieh and Natawidjaya, 2000). The dots (Fig. 2.2) depict inland earthquake events, whereas towards the ocean, earthquake sources are shown by Global Centroid Moment Tensor (CMT).

The giant Sumatra-Andaman earthquake that occurred on December 26, 2004 (Mw >9.0), which was followed by the Nias-Simeulue earthquake on March 28, 2005 (Mw=8.7) resulted in the northern part of Sumatra being the most affected area. The two major earthquakes, which hit in a short time interval and at near locations, released a large amount of seismic energy in the area. This phenomenon may also have contributed to the Padang earthquake.

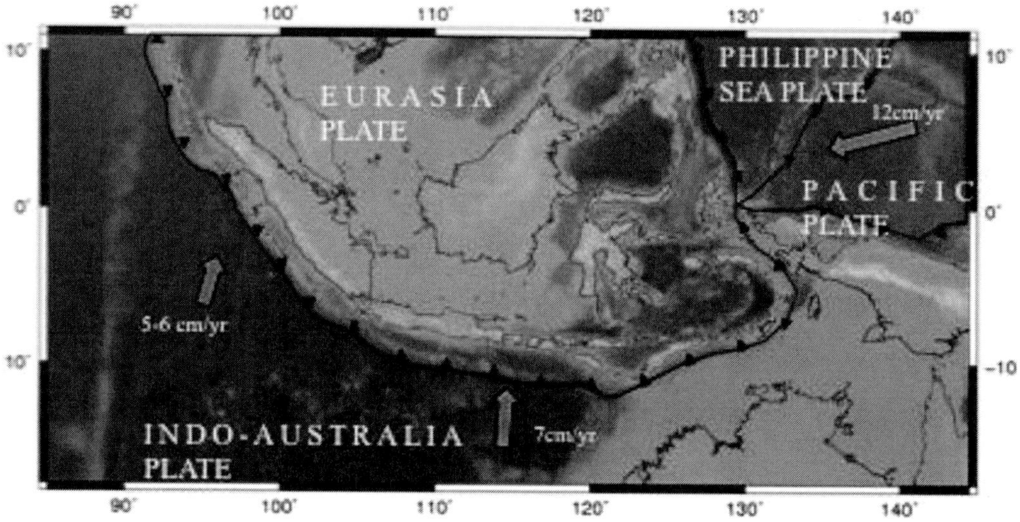

Figure 2.1. Active tectonic map of Indonesia (Source: Courtesy of Tsunami and Disaster Mitigation Research Center, Syiah Kuala University)

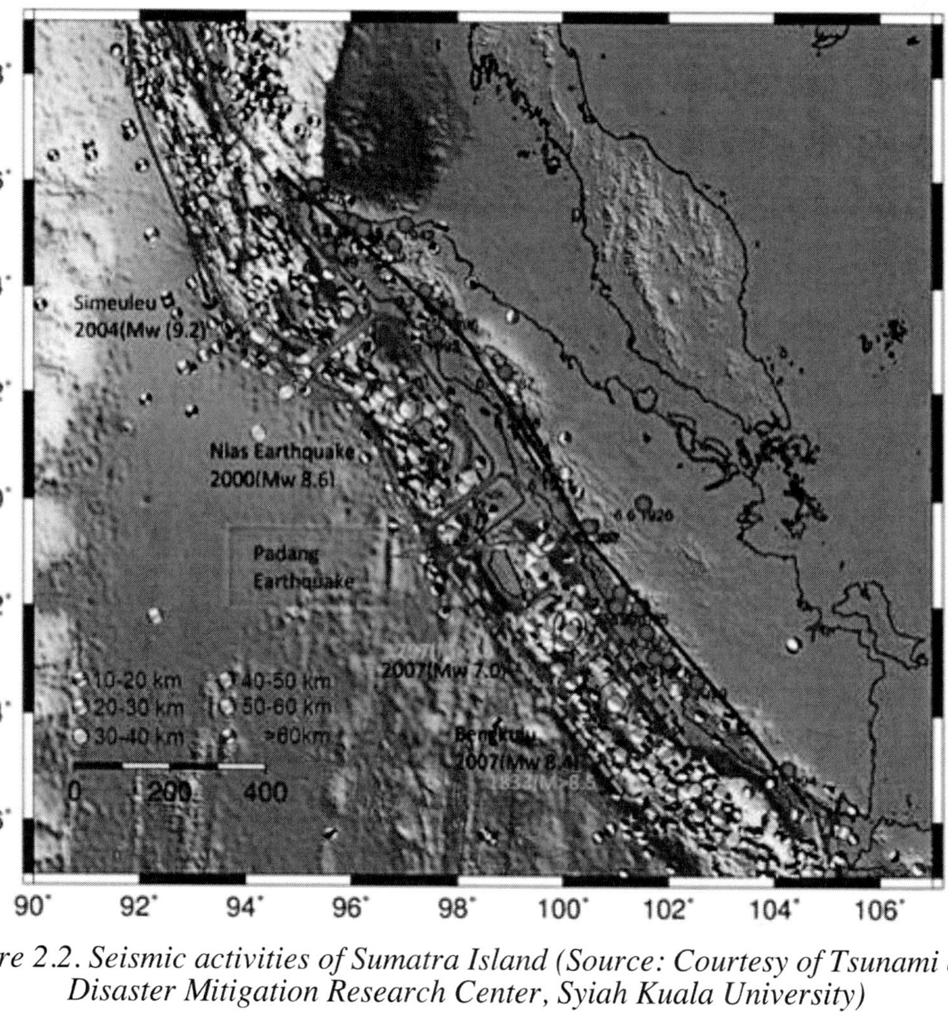

Figure 2.2. Seismic activities of Sumatra Island (Source: Courtesy of Tsunami and Disaster Mitigation Research Center, Syiah Kuala University)

If we observe the main structure of Sumatra more closely, it is oriented along a line running southeast-northwest starting from Sunda trench in the southern part of West Java, with the Bukit Barisan mountain range and Neogene foreland basin at the far end covering most of Northeast Sumatra. The most obvious topographical feature of Sumatra is its shear fault, which lies along Bukit Barisan mountain range and extends parallel to the trench and the line of active volcanoes along Sumatra. This is a shear fault that accommodates some of the shift from trench-parallel slip as a result of the Australian plate subduction along the western coast of Sumatra. The Australian Plate subduction velocity varies around 50 to 60 mm/year, from the east to the north, and consequently, it is not perpendicular to the direction of the trench.

Since 1890 there have been 21 earthquakes with magnitudes greater than 6.5 on the Richter scale along the Sumatran fault (Natawidjaya and Triyoso, 2007). Two of the greatest earthquakes have been the Liwa earthquake (Mw 6.4, in 1994) and the Kerinci earthquake (Mw 7.1, in 1995). Among the most recent earthquakes was the Solok earthquake (Mw 6.4 and Mw 6.3, on March 6, 2007). This along with others has resulted in categorizing the Sumatran fault one of the most active shear faults with the biggest influence on seismic activity in Southeast Asia.

Natawidjaja and Triyoso (2007) described the slip rate along the Great Sumatran fault consisting of 19 segments to be about 23 to 26 mm/year. In light of this, there are three major factors that could have contributed to the Padang Earthquake — the subduction zone, the Mentawai fault, and the main fault along Sumatra Island. These three effects are typical of a seismic catastrophe in Sumatra where every occurrence of submarine earthquakes (in the subduction zone) will be followed by an earthquake in the Sumatra fault system.

2.2 The Epicenter in the Subduction Zone

The Sumatran continental crust has a thickness of about 30 km (Malod and Kemal, 1996), while the depth of the seismic source (focus) of the Padang earthquake reached 79 km, meaning that the epicenter was below the slab (in the oceanic crust) resulting from subduction beneath the continental crust (Fig. 2.3). However, as force propagation occurs more easily through a fault in the vicinity, the epicenter of this earthquake may have emerged along the Mentawai fault.

There are other scenarios as well, in that the master fault, the Mentawai fault, as a regional dextral fault could be as a deep as Sumatran fault, which extends almost 100 km to the lower boundary of the lithospheric mantle (Malod and Kemal, 1996). Even if the Mentawai fault is only half the depth of the Sumatran fault, its profundity is approximately 50 km to the depth of the earthquake's epicenter about 20 km under the slab. Therefore, it is possible that at a depth of 79 km, the oceanic crust rheology is brittle enough to cause breakage of this crust contributing to Sumatran earthquakes.

It is possible that the existence of this Mentawai fault diverts the earthquakes toward Sumatra, and if there were no Mentawai fault, earthquake forces from a depth of 50 km would not affect Sumatra but would instead be diverted to the west of Siberut, toward the Indian Ocean.

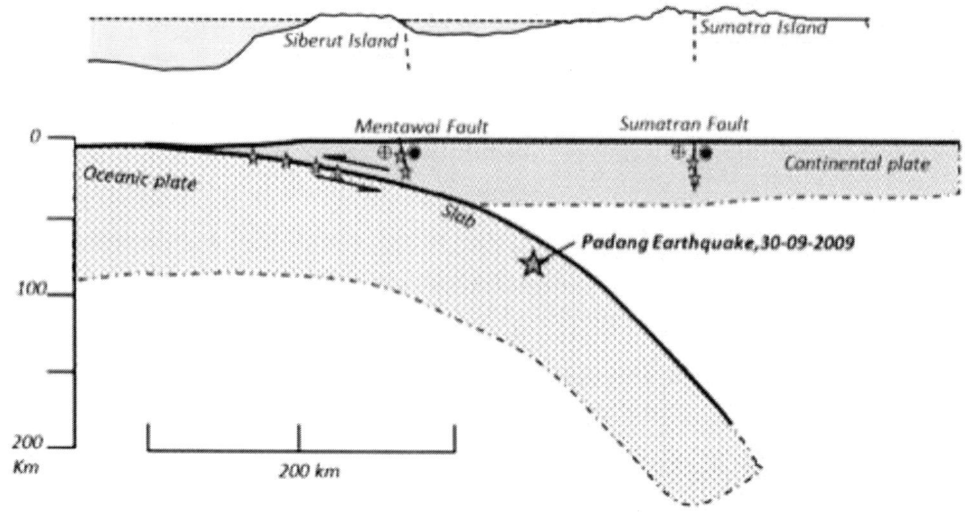

Figure 2.3. Cross-section of the fault system in Sumatra (Source: Courtesy of Tsunami and Disaster Mitigation Research Center, Syiah Kuala University)

2.3 Effect of Aftershocks

The occurrence of this giant earthquake caused an aftershock effect triggering another earthquake, the seismic segment and fault of which are close to the collapsed lithology of the source of the Padang earthquake. This can be observed in the subsequent earthquake that hit Jambi the next day as a result of Semangko fault.

2.4 Awareness and Mitigation

As mentioned above, the major concern of the Padang earthquake is that it was the result of the Mentawai fault. The seismic source was located at the western edge of the Mentawai segment (mega-thrust), which stretches from Siberut Island to Pagai Island. According to various studies on seismic history in Mentawai Island, there have been a number of giant earthquakes over the last hundred years. In general, the earthquake recurrence period on the island west of Sumatra is about 200 to 300 years, as a result of the Indo-Australian plate subduction by which the tip of the Eurasian plate is suppressed beneath the western side of the Semangko fault, extending to the islands on the coast of Sumatra. The velocity of the plate pressure is about 6 to 7 cm per year.

Due to its location in the surrounding faults instead of in the subduction zone, this earthquake is not considered a major earthquake in the 200-year cycle, which is currently believed to apply in the main zone of Mentawai segment. Thus in general, the occurrence of this earthquake could make the subduction segment more vulnerable. The earthquake that hit Padang did not further reduce potential energy in the Mentawai segment, but it could trigger a release of energy faster because the tension in the main zone is now higher due to direct interaction from both sides.

The Padang earthquake mechanism analyzed using Global Centroid Moment Tensor (CMT) showed that it was not of thrust-fault type as it is the case with earthquakes occurring in the inter-slab. Seismologists consider this earthquake an inter-plate (inside the plate) earthquake, which is moving under Sumatra (some experts call it "inter-slab" because it was located at a subducting slab). The thickness of the continental crust plate is approximately 30 to 50 km for Sumatra. Thus, it could be assumed that this earthquake did not occur in the contact area of the two plates. Due to the influence of temperature and pressure, at a depth of 80 km the interface that allows stick-slip from both rigid plates could have disappeared. Therefore, it is unlikely that this was an inter-plate earthquake.

In fact, the potential energy released in the Mentawai segment had been reduced following the Bengkulu earthquake on September 12, 2007, with magnitude of 8.5 on the Richter scale. The epicenter was in the same "pouch" of seismic energy, and the earthquake had reduced about a third of the total potential energy. It is expected that the energy left in the respective zones be gradually released. Based on historical data, earthquakes that occurred in 1650 and 1833 at this location caused devastating tsunamis, which reached approximately 10 m in height in Padang (Natawidjaja and Triyoso, 2007). It is an accurate assumption that the frequent earthquakes will be a positive means for the gradual release of energy from this fault; however, for Padang we should be aware of the potential for giant earthquakes from the Mentawai region, the vast amount seismic energy of which have not yet been released.

2.5 Acknowledgements

The original draft of this chapter was Mr. Didik Sugiyanto, MT of the Tsunami and Disaster Mitigation Research Center (TDMRC) of Syiah Kuala University, Banda Aceh, Indonesia. Dr. Teuku Alvisyahrin of TDMRC provided us with the translated version for the monograph. Their permission to use the figures is much appreciated.

2.6 References

Malod, J.A. and B.M. Kemal. 1996. The SumatraMargin: Oblique Subduction and Lateral Displacement of the Accretionary Prism, *In* R. Hall and D. Blundell. 1996. Tectonic Evolution of Southeast Asia.

McCaffrey, R. 1996. Partitioning Slope at Convergent Plate Boundaries of SoutheastAsia, *In* R. Hall and D. Blundell. 1996.Tectonic Evolution of Southeast Asia.

Natawidjaja, D.H., and W. Triyoso. 2007. The Sumatran Fault Zone-From Source to Hazard. Journal of Earthquake and Tsunami, Vol. 1, No. 1:21-47.

Sieh, K. and D. H. Natawidjaja. 2000. Neotectonics of the Sumatran fault, Indonesia, *Journal of Geophysical Research*, Vol. 105, No. B12, Pages 28,295-28,326.

Chapter 3
TRANSPORTATION

The transportation lifelines discussed in this chapter include roads and bridges. As in other Southeast Asia countries, the transportation infrastructure in Indonesia lags behind the proliferation of automobiles on the roads. Traffic congestion became more serious immediately following the earthquake even without a catastrophic failure of the transportation system.

The bridges performed reasonably well: there was no collapse. We were told that the bridges were designed to a high seismic standard that is equivalent to the United States. The most severe problem was abutment settlement and separation at the road junction. Twenty-nine bridges were reported damaged in the city of Padang. The most common road failure was asphalt surface cracks and ripples (Fig. 3.1). In Pariaman, landslides covered a number of country roads along the foothills (Fig. 3.2), delaying time-critical emergency response and rescue.

It was reported that the total cost of roads and expressway repairs was 70.5 billion Rupiah (US$7.55 million), and according to a report from the Pandang mayor's office, the costs to repair damaged bridges was 22.38 billion Rupiah (US$2.39 million).

During our investigation, the meeting was set up with the Traffic Division of the Transportation Department due to a translation miscommunication. However, this group was extremely helpful in pointing out the locations where damage occurred. The head of the Transportation Department responsible for the roads and bridges was uncooperative.

3.1 Description of System

The city of Padang and the greater Padang area are served by a network of expressways (Fig. 3.3). Two major expressways run north-south, connected by east-west expressways and city roads. The two north-south bound expressways are Jalan Hamka and Jalan By Pass. Jalan Hamka runs along the coastline; Jalan By Pass runs to its east. These two expressways join the Teluk Bayur Port on the south and close to Miangkubau International Airport on the north. The east-west expressways are near the city of Padang. The expressways are four-lane asphalt paved roads with some concrete sections; city and rural side roads are usually asphalt paved.

Figure 3.1. Surface cracks on roads

Figure 3.2. Landslides in Pariaman cover roads leading to small communities

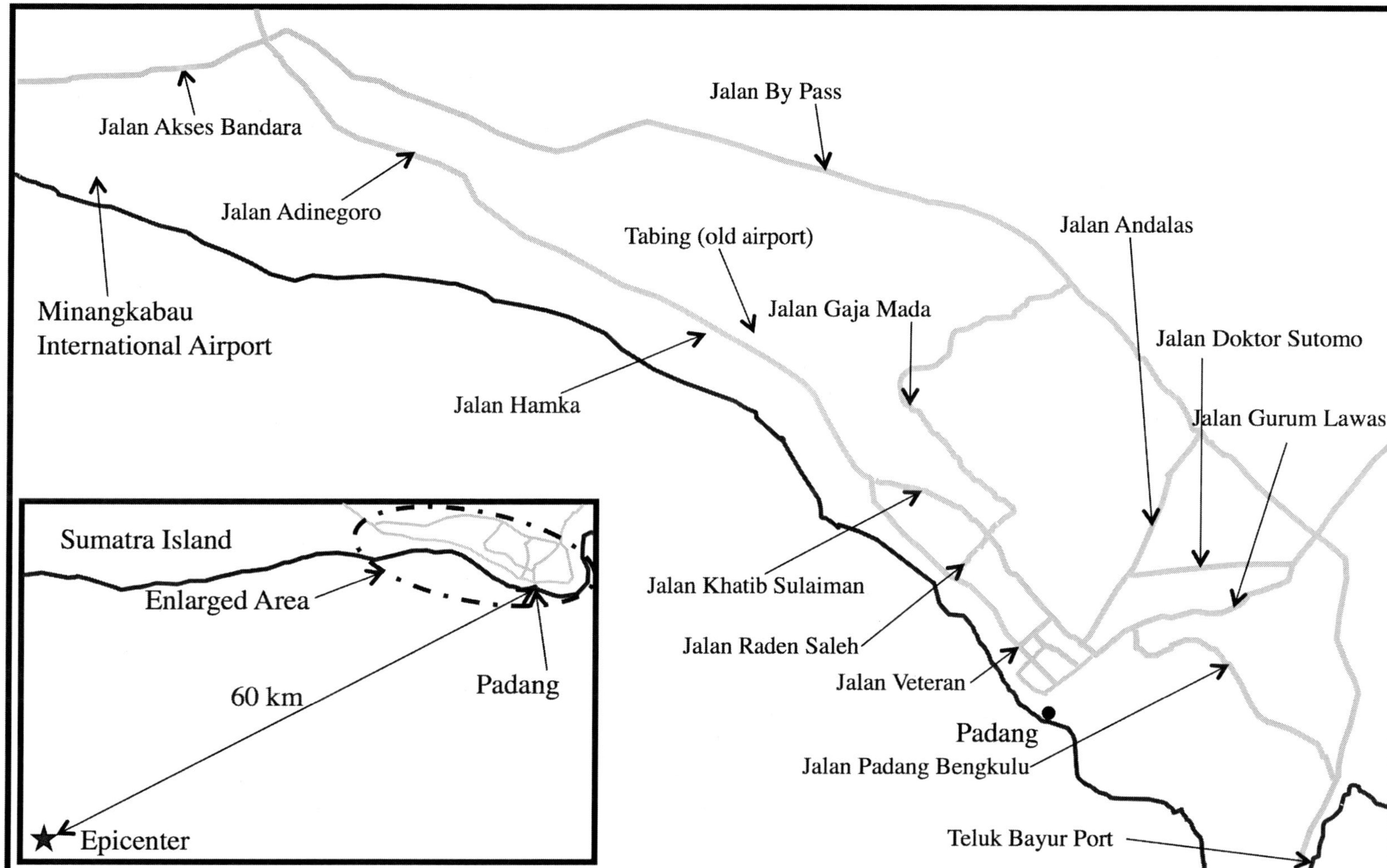

Figure 3.3. Road and expressway system in Padang (Source: modified from Google Earth Image)

Two major rivers (batangs) run east-west towards the ocean. Batang Arau is close to Padang, while Batang Kuranji is closer to Miangkubau International Airport. Banjir Kanal (Banjir Canal) runs between the two rivers. There are many tributaries joining these rivers and canal; as a result, there are many bridges in Pandang's road systems. Within the earthquake-impacted area, the transportation system has 38 road bridges and six railway bridges.

There is a single-track, dedicated freight railway running north-south between Jalan Hamka and Jalan By Pass. It is not for passenger trains. The track terminates at the Teluk Bayur Port.

3.2 Overview of System Performance

Overall, the performance of the roads and bridges was good. Emergency response and search rescue personnel experienced minor delays due to collapsed building debris, road surface cracks, and bridge abutments settlements. No bridges collapsed in West Sumatra Province. Thirty-one roads within the earthquake-impacted area were damaged. Most of the damage was surface cracks and ground settlements.

With massive landslides covering roads, emergency crews in Pariaman had difficulties in reaching victims. As most of the roads leading to the area were inaccessible, the team performed an aerial survey by helicopter, courtesy of the Mayor of City of Padang.

Of the 38 bridges within the earthquake-impacted area, only 29 had sustained damage. All the damage observed was confined to the abutment areas at the junction of the bridge and the road surface. A couple of the settlements were severe enough that cars had to slow down to go over the bumps, a result of temporary fixes by the Transportation Department. The Transportation Department staff indicated that five bridges were more severely damaged than the rest, and suggested we take a look. Two of the five bridges had colocated lifelines that were damaged.

None of the railway bridges, a total of six in the earthquake-impacted area, experienced any damage. The only damage we observed was settlement at the abutment of one railway bridge where sand bags were used to fill the void (Fig. 3.4) (S0.902919°, E100.352057°). As previously mentioned, all trains in the area are freight trains. There is one terminal for freight trains, and no passenger terminal within the city. During our 4 days in the area, we did not see any trains coming through.

3.3 Bridge Damage Site 1 (S0.95966°, E100.40419°)

The bridge at Damage Site 1 was one of five damaged bridges we visited. It is located on the south end of the Jalan By Pass. Following the Transportation Department /Traffics Division instructions, this was the first damaged bridge we saw along the Jalan By Pass traveling northbound. Figure 3.5 shows the locations of the five bridges the TCLEE post-earthquake investigation team visited. The bridge at Bridge Damage Site 1 (identified on Fig. 3.5 as Bridge 1) is approximately 100 m long and two lanes wide (Fig. 3.6). Each lane seems to be wider than needed for one car; the extra space is for motorcycles and motor tri-cycles. This is a single-span bridge with one pier supporting the concrete I-bean shape girders in the middle. The deck is concrete covered with asphalt. The ends sit on the 40-cm-wide shoulder of the concrete abutments.

Figure 3.4. Sand bags fill the void created by slumping on the south end of this railway bridge.

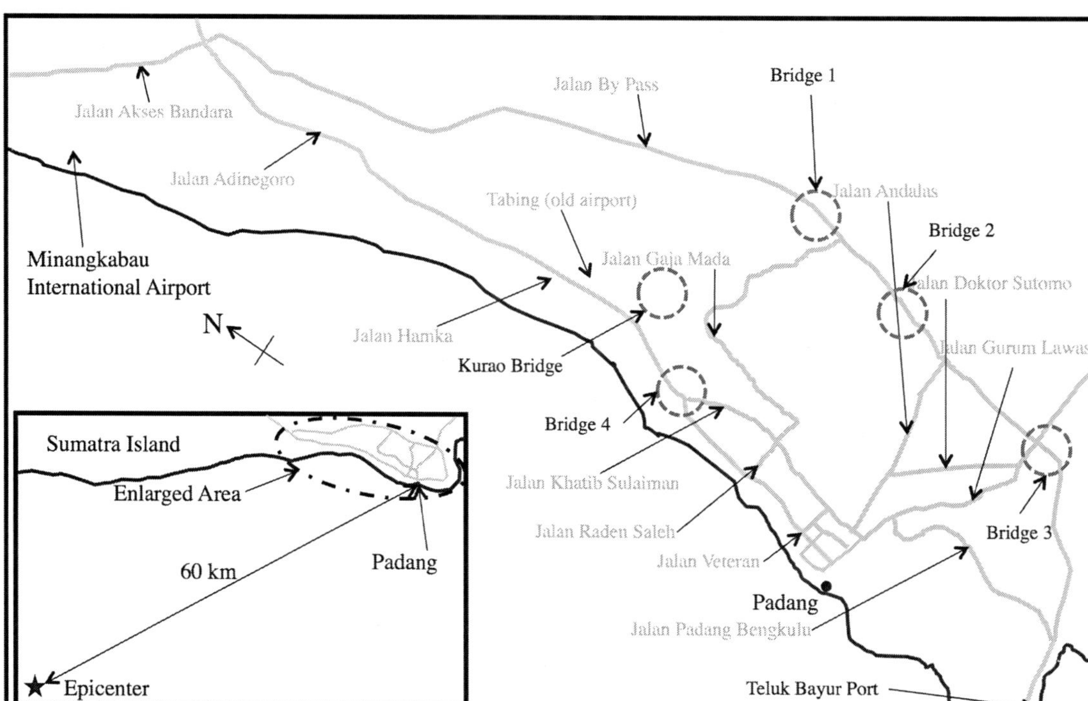

Figure 3.5. Locations of the five damaged bridges visited by TCLEE team. Bridge Damage Site 1 is the circle on the far right.

Figure 3.6. The first bridge visited on the Jalan By Pass that showed light damage. The photo was taken from the north end of the bridge facing south.

Damaged Observed at (S0.95966°, E100.40419°)

The team observed light damage at this site on the north end of the bridge. In a few places, there was spalling concrete around the railing posts along the side. Figure 3.7 shows typical damage observed. Most of the damaged sites were not repaired. See the Executive Summary for repair estimates. The costs of temporary fixes are also included in the overall loss estimate.

On the west side of the bridge, there are a couple of colocating lifeline conduits (Fig. 3.8), which are most likely telecommunication cable conduits. There were also a couple of holes on the concrete block of the abutment, which were probably prepared for lifelines such as electric power and telecommunication cables (see the lower left side of Fig. 3.6).

The road surface on the north end of the bridge subsided and was temporarily repaired with asphalt (Fig. 3.9). This will most likely be the permanent fix. The south end of the bridge did not show any sign of damage (Fig. 3.10).

This section of the Jalan By Pass is heavily travelled in both directions. According to the Transportation Department, the bridge damage was repaired the day after the earthquake.

Figure 3.7. Except for the road surface, the damage shown was not fixed. There were four locations with similar damage on this bridge.

Figure 3.8. Two conduits for lifelines were mounted on the west side of the bridge. Note the barbwire wrapped around the conduits to prevent people from walking or climbing on them.

Figure 3.9. North end of the bridge at the abutment area subsided and was temporarily fixed.

Figure 3.10. There was no sign of damage on the south end of the bridge.

3.4 Bridge Damage Site 2 (S0.92190°, E100.39762°)

The second bridge visited was northbound on Jalan By Pass (see Fig. 3.5 for location). This bridge showed a little more significant damage than the first. The bridge is about 140 m long and two lanes wide. The typical lane width is about 1.5 lanes, with the extra space intended for motorcycles and motor tri-cycles. It is a single span bridge with I-beam shape concrete girders and concrete deck supported by one pier in the middle (Fig. 3.11). The ends rest on the approximately 40-cm-deep shoulders of the concrete abutments (Fig. 3.12).

Damage Observed at (S0.92190°, E100.39762°)

The concrete spalling indicates that the bridge was hinged on the south end of the concrete deck (Fig. 3.12). The north end was displaced about 20 cm to the east, indicating a clockwise rotation (Fig. 3.13, 3.14). There was no sign of movement of the girder's bent cap.

In addition to the shifting of the bridge deck, there were a few places of ground settlement and surface cracks on both ends of the abutment to road junction areas. These areas were temporarily fixed with asphalt (Fig. 3.15).

Similar to the first bridge, this one also has lifelines embedded in the concrete deck as well as mounted on its side (Fig. 3.16, 3.17). Because the movement was not excessive, the conduit was not damaged. However, the anchor bolts that secured the metal bracket to the deck were sheared, and part of the bracket was broken (Fig. 3.17).

Figure 3.11. The second bridge studied along the Jalan By Pass

Figure 3.12. Abutment details on the south end of the bridge

Figure 3.13. Damage to Bridge 2

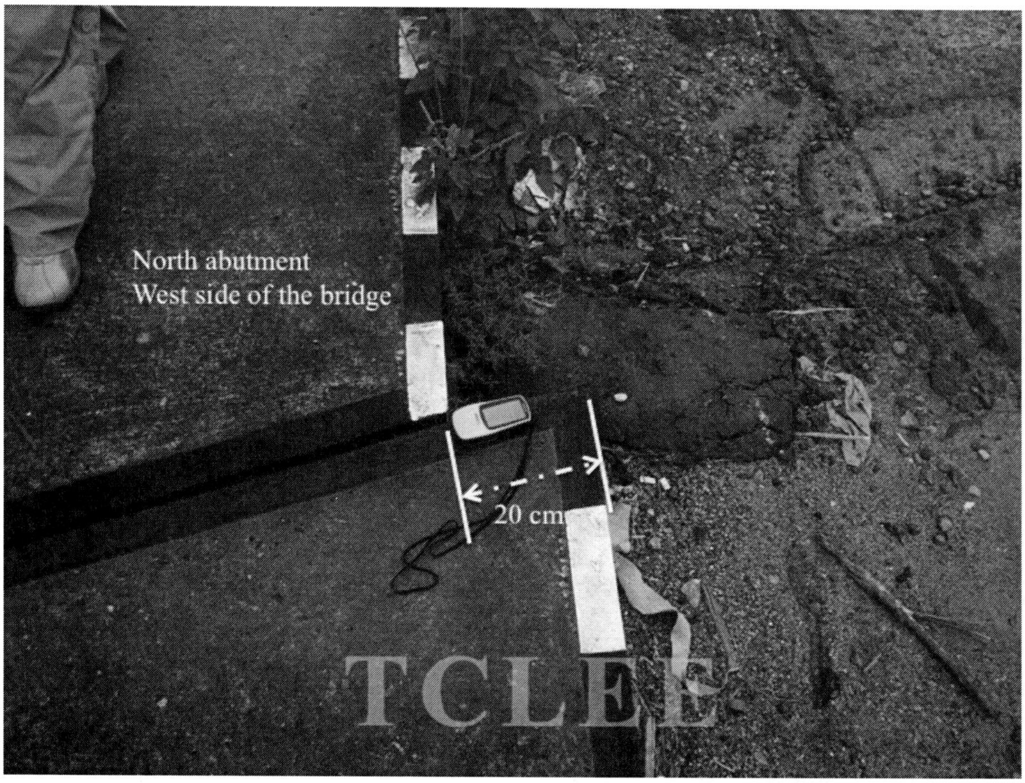

Figure 3.14. Bridge deck shifted about 20 cm to the East

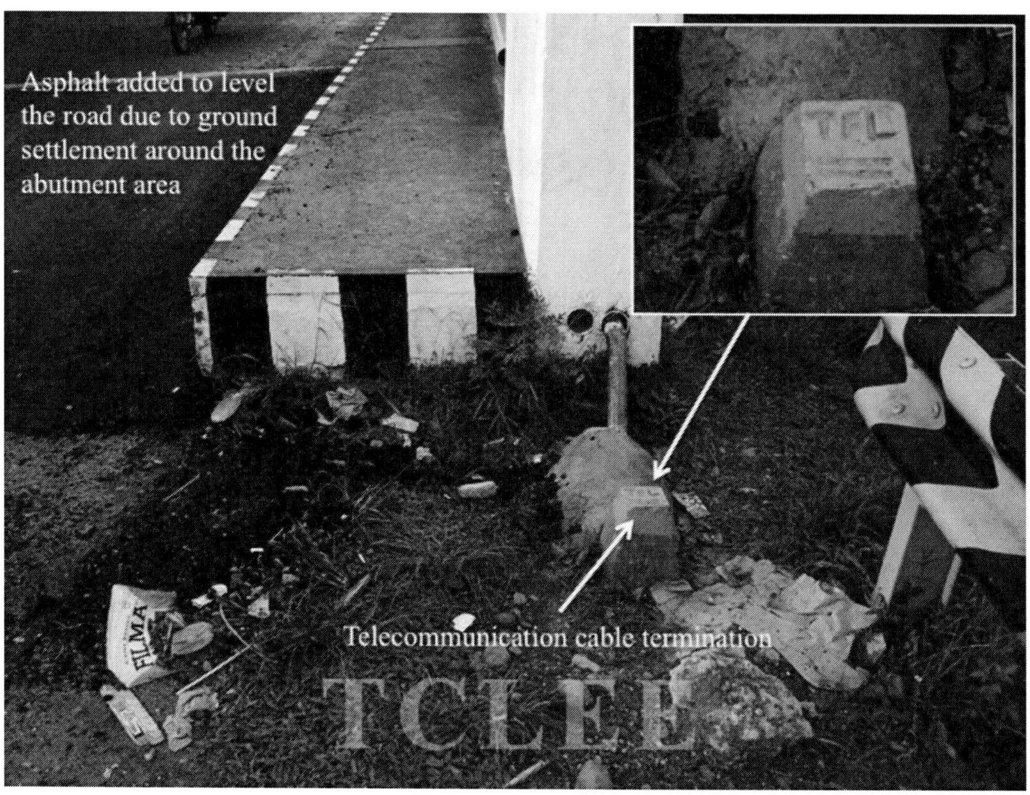

Figure 3.15. Telecommunication cables routed inside the conduit embedded in the bridge deck

Figure 3.16. Colocating lifelines secured along the bridge

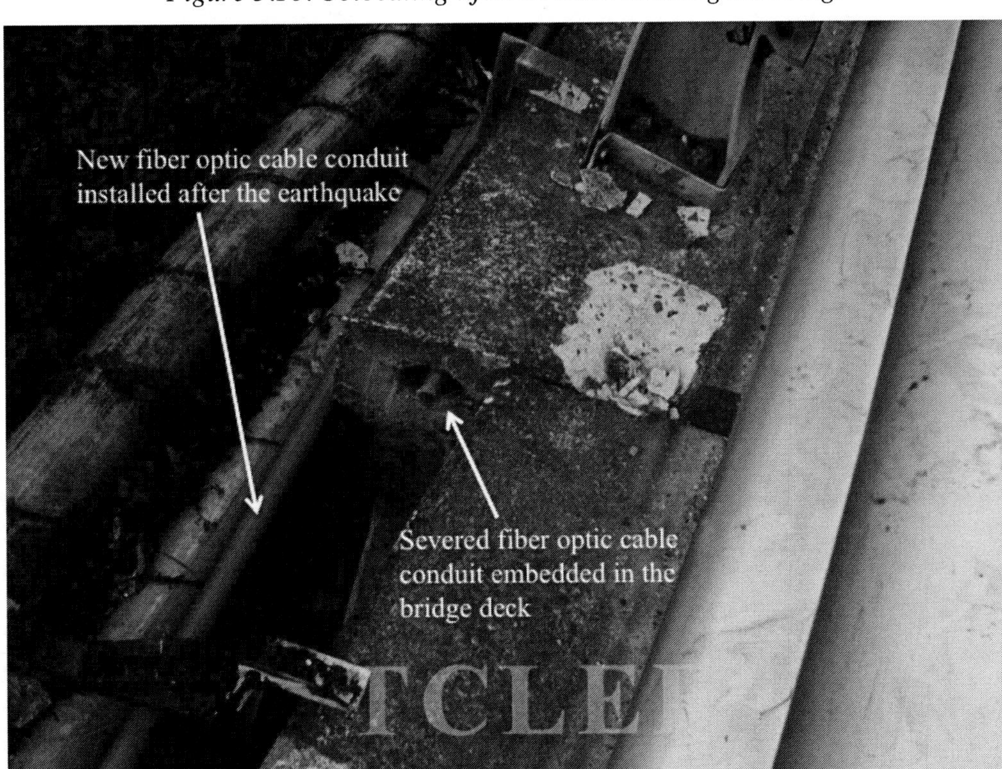

Figure 3.17. Severed telecommunication cable embedded in the bridge deck with new cable installed to restore service

3.5 Bridge Damage Site 3 (S0.89264°, E100.395039°)

Traveling northbound along the Jalan By Pass, the team encountered the last of the three damaged bridges on the road identified by the Transportation Department. This bridge is shorter than the two discussed in Sections 3.3 and 3.4. It is two lanes wide and about 50 m long (Fig. 3.18). The design and construction are similar to the other two bridges discussed. It also carries lifelines along its west side and in the conduits of its deck (Fig. 3.24). The metal conduits were casted in place when the deck was formed.

Summary of Damage at (S0.89264°, E100.395039°)

A lateral movement of the bridge was evident at the expansion join along its north end (Fig. 3.22). The gap on the west side is about 60 mm wide but less than 55 mm on the east side, indicating that there was a slight counterclockwise rotation of the bridge. The excessive gap on the west side exposed the optical fiber conduit running inside the embedded conduit in the bridge deck. The south end of the bridge shifted to the east relative to the abutment, with a displacement of about 120 mm (Fig. 3.18). Because it was dangerous to descend to the bottom of the bridge, the team assessed the cause of the offset from the top and concluded that the abutment shifted relative to the bridge during strong ground shaking (Fig. 3.19). The bridge also had several surface cracks where some ground settlement was also noted (Fig. 3.20). The lifelines (telecommunication cable and

Figure 3.18 Overview of the bridge facing north (S0.89264°, E100.395039°)

Figure 3.19. The offset at the south end of the bridge as seen from the east side

Figure 3.20. The north end of the bridge. Note the longitudinal crack on road surface close to the abutment.

power cable) were mounted on the west side of the bridge. The two conduits with barbed wired wrapped around them were secured on metal brackets anchored to the bridge deck by pipe cramps (Fig. 3.23). The embedded conduit for optical fiber cable is on the west side of the bridge (Fig. 3.24). Surface cracks on this bridge were not repaired because they were not in the path of the traffic (Fig. 3.21). Ground settlement was also evident northwest of the abutment.

On the east side of the bridge, brackets were installed and the embedded conduit was also in place for colocating lifelines (Fig. 3.25). Fortunately, there was no lifeline installed in the embedded conduit; otherwise, it would have been severed. Colocating lifelines along bridges and overpasses of highways is a common practice globally. Collateral damage has been observed in all post-earthquakes investigations by TCLEE.

Figure 3.21. Surface crack on the north end east side of the abutment of the bridge

Figure 3.22. The expansion joint opened up to about 60 mm on the west side.

Figure 3.23. On the northwest side of the bridge, the lifelines colocating with the bridge were exposed. The two embedded conduits were for telecommunication cables.

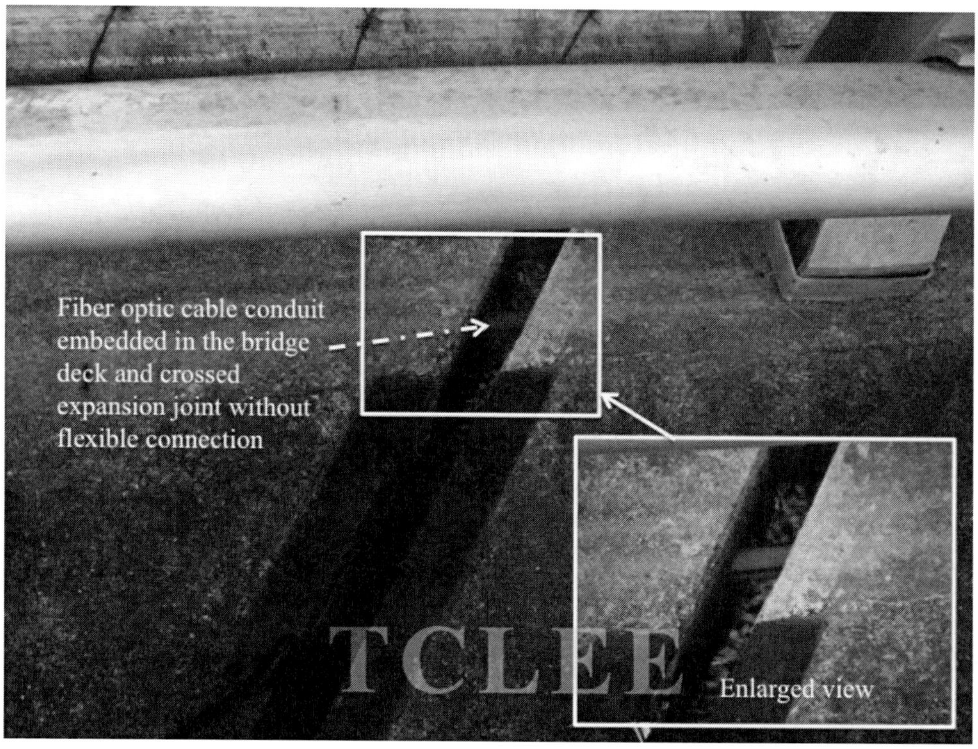

Figure 3.24. The gap at the expansion joint exposed the optical fiber conduit inside the embedded conduit in the bridge deck.

Figure 3.25. East side of the bridge at (S0.89264°, E100.395039°). Note the brackets and the hole (lower left of this figure) prepared for routing lifelines.

Figure 3.26. The two bridges on Jalan S Parman. On the left (west) is the concrete girder bridge, and the right (east) is the steel girder bridge.

3.6 Bridge Damage at Minang Plaza (S0.903533°, E100.350845°)

One span of the twin bridge, Jalan S Parman, at Minang Plaza is for southbound traffic, the other for northbound traffic. The plaza is on the north end of these bridges. The span for the northbound traffic was lightly damaged. Both bridges are about 110 m long and two lanes wide (about 8 m including pedestrian walkways). Just east of the northbound traffic bridge and about 160 m on the south side of the river is a steel truss railway bridge. There was no damage to this bridge (Fig. 3.1).

Both bridges have a concrete deck with girders supported by bents. The southbound bridge has precast I girders, while the northbound bridge has steel I girders. Figure 3.26 shows the two bridges; on the left is the southbound traffic bridge. This bridge does not have lifelines colocating on it. The bridge for northbound traffic (Fig. 3.27) has a transmission water pipeline mounted on steel T-bar (inverted) under the deck and between the steel I-beam girders on the east side. The steel pipeline is about 180 mm in diameter (OD) and emerges from the ground on the southeast side of the bridge (Fig. 3.28) then routes across the river under the bridge (Fig 3.28). There was no evidence of any damage to this pipeline.

Figure 3.27. The bridge on the east side with Ninang Plaza on the right

Figure 3.28. Water transmission pipeline under the bridge

Figure 3.29 A damaged concrete guardrail post on the west side of the bridge

Figure 3.30 A damaged concrete guardrail post on the east side of the bridge

Summary of Damage at (S0.903533°, E100.350845°)

The observed damage to this bridge is minor. There were three locations of damaged concrete guardrail posts on both sides of the bridge and minor damage to the pedestrian walkway on the east side (Fig. 3.29, 3.30). The expansion joint metal cover plate popped out of its seat (Fig. 3.29). There was also slight abutment wall crack on the northeast of the bridge. All the damaged posts are located at the expansion joints.

3.7 Kurao Bridge (S0.888281°, E100.369829°)

This was the last bridge site the team visited in Padang. This bridge is located about 2.7 km northeast of the twin bridge at Ninang Plaza. The design and construction of this bridge is the same as the bridges described in Sections 3.3. 3.4, and 3.5. The bridge is about 60 m long and wide enough for two-lane traffic (Fig. 3.31).

This bridge has a steel water transmission pipeline routed on its east side (Fig. 3.32). The road surface on its south end showed significant damage, and the wingwall wall was partially broken (Fig. 3.32). The pipeline underneath was secured by a series of metal brackets that hung from the concrete guardrail posts (Fig. 3.33). A major pipe break occurred at this location. Most of the damage was observed on the south end of the bridge.

Summary of Damage at (S0.888281°, E100.369829°)

At the south end of the bridge, the damaged road surface was not repaired (Fig. 3.34).

Figure 3.35 shows the damage at the west abutment. The concrete walkway was dropped about 100 mm. It had no reinforcement or temperature bars and was not connected to the abutment. The top of the staircase (Fig. 3.36) was also dislocated from the abutment, and part of it was broken. Gravel was used to fill the voids around the abutment.

Figure 3.31. Kurao bridge looking south

Figure 3.32. The broken wingwall showing the water pipe on the east side of the bridge

Figure 3.33. Metal bracket securing the steel pipe under the bridge

Figure 3.34. The abutment sidewalks were slanting at about 20°

Figure 3.35. The concrete sidewalk pad dropped and moved away from the abutment.

Figure 3.36. The staircase leading to the ravine was damaged and dropped below the walkway.

3.8 Road Damage in Padang (S0.938032 , E100.351261)

At a few locations in Pandang, the road surface sustained both lateral and transverse cracks. The most severe damage was to the road along the coast, which was closed to traffic during TCLEE team visit.

We were fortunate to spot the damage during our helicopter ride, and photographed it after we landed. The location is north of the intersection of Jalan Samudra and Jalan Purus 3 on Jalan Samudra. Figure 3.39 shows the extent of the surface crack from the air, and Figure 3.40 shows the width and depth of the surface crack. These cracks were a result of lateral spreading toward the sea. The longest crack was about 50 m. There were a few locations showing sand boil due to liquefaction.

On the east side, the concrete sidewalk pad moved away from the joint between the bridge deck and the abutment (Fig. 3.37). The broken steel pipe was repaired (Fig. 3.38).

Figure 3.37. The south abutment sidewalk pad on the east side separated from the bridge deck.

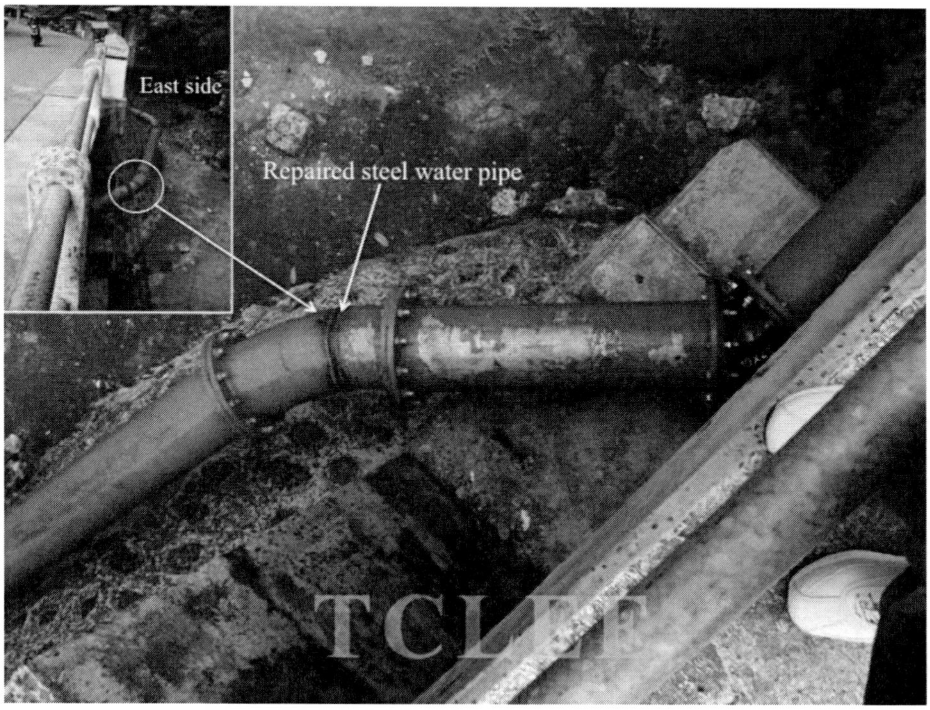

Figure 3.38. The broken pipe was welded.

Figure 3.39. Aerial photo of the road surface crack

Figure 3.40. The road surface crack as seen facing north.
See the circle to the right of Fig. 3.39 for location of the cracks.

3.9 Landslides in Pariaman

The numerous landslides, both above the road and down its slope side that provides access to residential areas in the mountainous region of Pariaman, made travel in this area extremely difficult. With loss of electric power and telecommunication, the local population became anxious quickly. Fortunately, this area is not rock filled. Getting around was very difficult, but we overcame it with light equipment.

These areas essentially had no slope management. It is evident from the photos that there were no retaining walls or water drainage on either the cut side or the down slope side. Heavy seasonal rainfall coupled with the big earthquake easily triggered landslides in the area (Fig. 3.41 to 3.46). Residents here were trapped, and their houses were completely or partially destroyed by the earthquake. Some roads became so narrow due to down slope failure that cars could not get through (Fig. 3.46).

Figure 3.41. Landslide blocking a portion of the road leading to a small community up the hill

Figure 3.42. A landslide covering a couple of kilometers of road

Figure 3.43. Downslope landslide threaten the road above

Figure 3.44. Another downslope landslide narrowed the road

Figure 3.45. A downslope landslide narrowed the road to this community where several houses collapse.

Figure 3.46. Another view of the location in Figure 3.43; a house slid with the downslope landslide.

3.10 Major Observations and Recommendations

No bridges collapsed in Pandang; thus, the performance of the roads and bridges was very good. Although we were unable to obtain the bridge design documents, from our observation the bridges were well designed and constructed. However, the Public Works Department should evaluate the bridge structures and compared them to the most current code and standard available to ensure they can withstand the bigger earthquakes expected in this region.

The utilities, such as telecommunications, electric power, and water, should develop and implement redundant routing to prevent future loss due to colocating with bridges.

In light of the landslides in Pariaman cutting off access to the communities in the hills, alternate routes shall be developed to allow residents to leave or get emergency response quickly. Better slope management to reduce landslides is an alternative remedy to keep roads open after an earthquake.

3.11 Acknowledgements

The author is indebted to Professor HS Oey of University of Texas, El Paso, for translating documents from Javanese to English. Our contact, Mr. Eri Arsya, also set up a meeting for us with the Public Works Department. Additionally, Mr. Henky Mayaguezz provided documents containing financial loss data of lifelines. We thank these individuals for their significant contributions to our report.

Chapter 4
ELECTRIC POWER

The electric power transmission and distribution in West Sumatra is managed by PT Perusahaan Listrik Negara (PLN), a state-owned entity. Household power has two voltages, 127 V and 220 V both at 50 Hz. Kota Padang and Kota Pariaman are the two districts that suffered extensive low-voltage transmission network damage. About 20 percent of the customers in these districts lost power, and many outages lasted as long as 3 days after the earthquake. It is expected the power needed for these districts will remain low until they are rebuilt. The damage was concentrated in the distribution system, particularly the drop wire from poles to buildings. There were no large substations (500 KV) or power generation plants in the earthquake-affected area. Therefore, the damage to the electric power supply was not considered significant by a local report. The 20 KV low-voltage network sustained the most damage from severed cables, fallen poles, and damaged pole mount transformers (substations).

However, from a lifeline interdependence point of view, the loss of power affected the telecommunication sector and gas distribution outlets for as long as 3 days in some areas. This is critical for emergency response and recovery efforts.

Indonesia has had many blackout and brownout problems from power generation shortage and high demand over the years. Many businesses, such as hotels, and the telecommunication industry had installed electric power generating capability on their premises to reduce loss due to commercial power outages. The improved power supply reliability has had a negative effect on the power generating units owned by private businesses. Because these units are now seldom used, owners have stopped maintaining them properly, thus creating problems when they are needed during emergencies.

For wireless (cellular) service providers, the reliable power reduced the need for more reserved power (batteries) to sustain longer cellular base station operation in the case of power outage. In this earthquake, more than 20 percent of the base transceiver stations (BTS) went down due to power loss, which combined with physical damage to BTS, impaired the cellular network for more than a day after the earthquake. Small mobile power generating units had to be sent from Jakarta to Padang to restore BTS operation.

The total loss to PLN was about 52 billion Indonesian Rupiah (Rp) (US$5.56 million) that included a Rp 6 billion (US$ 640,000) loss of revenue.

4.1 Description of System

The electric power source of West Sumatra province comes from PLN and one small independent power producer (IPP). The total generating capacity is 520 MW, with 60 percent from hydropower and the remaining from fossil fuel. The power output is much

Figure 4.1. A substation, most likely a 230 KV substation, north of Pariaman showed no damage.

less than the capacity due to recent draught, and PLN has had to redirect power from both South Sumatra and Jambi grids.

The primary transmission network is 500 KV between major cities within Indonesia, supplemented with a 230 KV and 120 KV transmission network and then down to a 20 KV distribution network. The transformers of the 20 KV network are pole-mounted.

In Padang and Pariaman, the two districts in the province affected by the earthquake, there were no 230 KV or 120 KV substations. The team did see one substation during the helicopter tour to the landslide areas to the north of Pariaman. No damage was observed (Fig. 4.1).

4.2 System Performance

It took 10 days before the power was fully restored to earthquake-impacted areas. However, power was restored to businesses and critical facilities within 3 days of the earthquake.

Distribution System

The 20 KV substations in the low-voltage distribution network did not perform well. PLN reported that 1870 of the 20 KV transformers (substations) were damaged, and 176 transformers had fallen from their pole mounts. As the 20 KV transformers are easy to install, installing and repairing them did not take much time or require much equipment or technical staff. Figure 4.2 shows a typical 20 KV transformer installation. The transformer is not anchored; it rests on two U-channels, which are part of the base of the platform. A damaged 20 KV transformer is shown in Figures 4.3 and 4.4. Figure 4.5

shows two repairmen installing a new transformer on the platform mounted between the utility poles. The transformer is seated on a platform that provides extended space for a serviceman to work on it. New transformers were noted in many locations in Padang (Fig. 4.6).

The low-voltage distribution network cables were severed in many locations due to collapsed or partially collapsed building (Fig. 4.7, 4.8). We saw this all over the city wherever there was a collapsed or partially collapsed building.

A program to improve cable management should be initiated to reduce recovery interval when there is damage at locations such as shown in Figure 4.9

Some control buildings in substations sustained minor damage, but it did not affect the substation equipment or operation. Because these locations are remote, the TCLEE team was unable to visit them; however, we did observe a substation during our helicopter tour of the landslide areas (Fig. 4.1).

Figure 4.2. The transformer is located in between the utility poles about 5 m above ground.

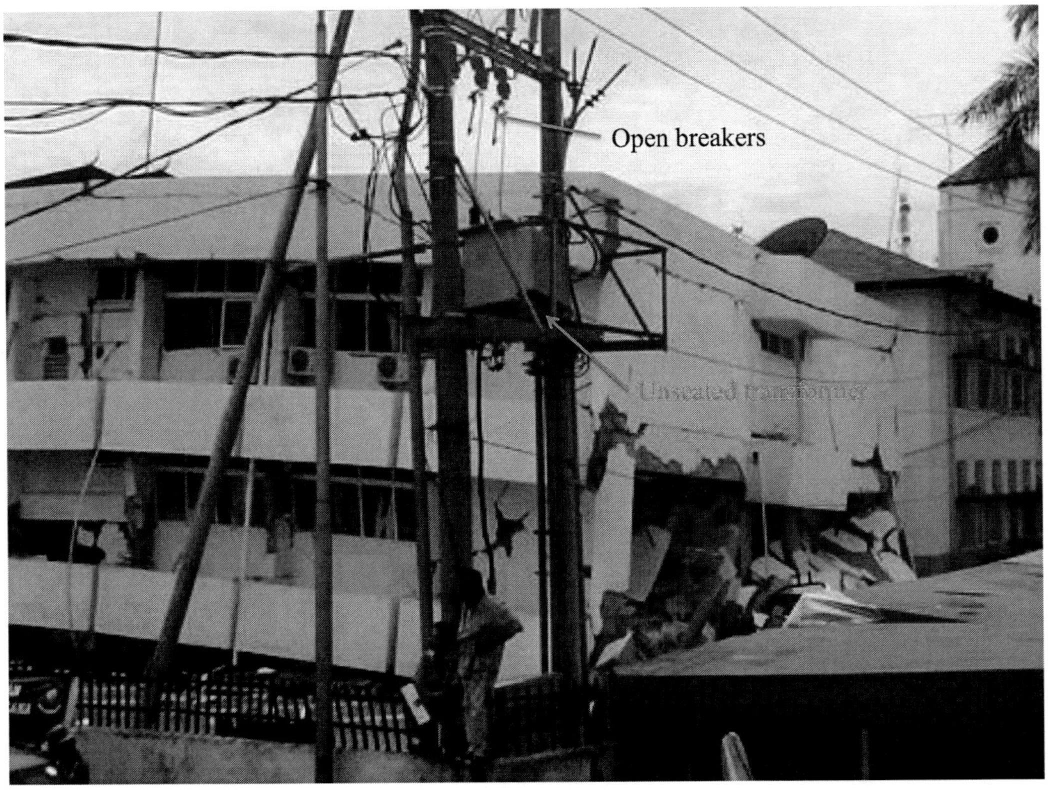

Figure 4.3. A damaged pole mount transformer in the 20 KV circuit. (Courtesy of Padang Fire Department)

Figure 4.4. A damaged 20 KV transformer with opened breaker above the transformer (Courtesy of Padang Fire Department)

Figure 4.5. Two service men were installing a new 20 KV transformer

Figure 4.6. A newly installed 20 KV transformer

Figure 4.7. Severed power cable due to collapsed building (Courtesy of Padang Fire Department)

Figure 4.8. Severed drop cables (Courtesy of Padang Fire Department)

Figure 4.9. Typical location where improved cable management will reduce recovery time

Transmission System

PLN reported that the high- and medium-voltage circuits were not affected by the earthquake. Figure 4.10 shows an undamaged high-voltage transmission tower in northern part of Pariaman district that is close to the landslide area.

Backup Power Generation

The emergency response and rescue units brought mobile power generating units to operate power tools for the search and rescue effort (Fig. 4.11).

Fire Following Earthquake

The fire department reported that most of the fires in the city following the earthquake were started by sparks from severed electric cables that were still energized.

There is a need to evaluate the platform design and the anchoring of the 20 KV transformer. Strengthening the structure and anchoring the transformer to it will significantly reduce damage in future earthquakes.

Figure 4.10. A high-voltage transmission tower that was not affected by the earthquake

Figure 4.11. Mobile power generator used by the Fire Department to operate power tool to gain access to collapsed or partially collapsed structures
(Courtesy of Padang Fire Department)

4.3 Major Observations and Recommendations

A severed drop cable from a building collapse can have the cascading effect of pulling down the aerial cable circuit. Feeding power using underground cable conduits to the buildings can reduce risk of that failure mode. Better cable management of the low-voltage transmission circuit will greatly improve the recovery performance as well.

A study by the U.S. Federal Emergency Management Agency (FEMA) shows that the cost-benefit of mitigation is about 1 to 4. Thus, one unit of spending typically results in four units of savings/efficiencies. Earthquakes as well as other disasters invariably teach us to mitigate to reduce loss.

4.4 Acknowledgements

The author greatly appreciates the information and photos provided by the Padang Fire Department. The author also would like to thank Hendro, who acted as our translator during our meeting and continued to help through e-mails afterwards.

Chapter 5
WATER SYSTEMS

5.1 Description of System

The Padang water system is served through three concrete water treatment plants with a production capacity of 300, 500, and 700 liters per second (l/s). These plants serve the north, south, and central areas of Padang. The network includes 1800 km of steel, asbestos concrete pipe (ACP), galvanized steel pipe, ductile iron pipe, and polyvinyl chloride (PVC), with welded or spigot joints enforced with rubber gaskets.

The Padang water system, which is operated manually, has more than 23,841 m in transmission lines and covers more than 65,000 household and commercial connections. It has about 400 employees, 12 cars, and two water tanks. Most pipe connections use a bell and spigot joint system with rubber gaskets.

Northern Area

The northern area with more than 21,000 connections is served from the Lubuk Minturun and Kuranji water treatment plant and three deep wells with the production capacity of about 300 l/s. Almost half the distribution system was damaged by the earthquake. The water pressure is generally very low (0.5–1 atm[1]) and, therefore, there was no treat of major eruption. Almost 1 month after the earthquake, 60 percent of the households had running water again. The remaining homes were served by mobile water tanks.

Central Area

The water source for the central area is from the Kuranji River. The Gunung Pangilun water plant has a capacity of 500 l/s. The primary distribution system is thorough ductile iron pipes, of which the primary pipe is 600 mm. The central area of Padang has more than 26,000 connections.

This area of Padang had the worst earthquake damage. After the quake, there were uneven settlements of up to 20 to 30 cm in the water plant's generator room. The 600 mm diameter pipeline had major damage and was repaired one week after the earthquake. One month after the quake, water distribution was restored to about 50 to 60 percent of the households. The Red Cross and other non-governmental organizations (NGOs) provided water for the remaining population through mobile tanks that visited neighborhoods on a rotating basis.

Southern Area System

The water source for the southern area of Padang is the Sikayan Balamuik and Arau rivers, which are fed by several creeks along the way. This water is then treated in the Ulu Gadut water plant with the production capacity of 220 l/s and saved in the Selatan

---

[1] 1 atm = 14.7 psi

reservoir. There are also several wells that are more than 46 m (150 ft.) deep. Additionally, water from the mountains is lightly treated and distributed.

The water distribution system for this area is all gravity flow through galvanized iron pipes, which change to PVC (150 mm, 200 mm, and 250 mm). The system serves more than 17,000 households.

There was no serious damage to the plant or reservoir during the earthquake. However, the heavy rains that followed caused landslides on October 24, 2009, damaging the water supply lines for this area. Many water lines were severed. In our helicopter survey of the landslide area, we could see water pipes scattered at the bottom of slopes (Fig. 5.4). Figure 5.5 shows the pipes at ground level scattered along the riverbank. There was minor damage at the treatment plant.

For emergency water supply, the water level in the river was raised using temporary dams, which also increased well intake. These initiatives provided water to 50 percent of the households in southern Padang three days after the landslides.

Repair materials such as those shown in Figure 5.6 were shipped in from various locations. There was no shortage of material. Many small leaks were difficult to locate due to heavy rains after the earthquake. During our visit, we could spot many areas with small leaks and puddles on the ground (Fig. 5.7).

5.2 Water System Performance

The water system was one of the lifelines severely impacted in this earthquake. Potable water was a problem; however, there was no shortage of water due to the rainy season.

As mentioned, the pipeline system suffered minor damage as the result of rainfall that triggered landslides in the foothills just south of Pandang. More steel pipes than PVC pipes were damaged in this earthquake. Figures 5.8 and 5.9 show the repair to damaged steel pipes.

There was also damage at the Sumur Bor-3B water treatment plant. The steel intake pipe was damaged and repaired by welding the parts back together (Fig. 5.10). The building sustained some damage as well (Fig. 5.11), while the pipes to the pump in the pump house sustained minor damage (Fig. 5.12). These problems were resolved within a week.

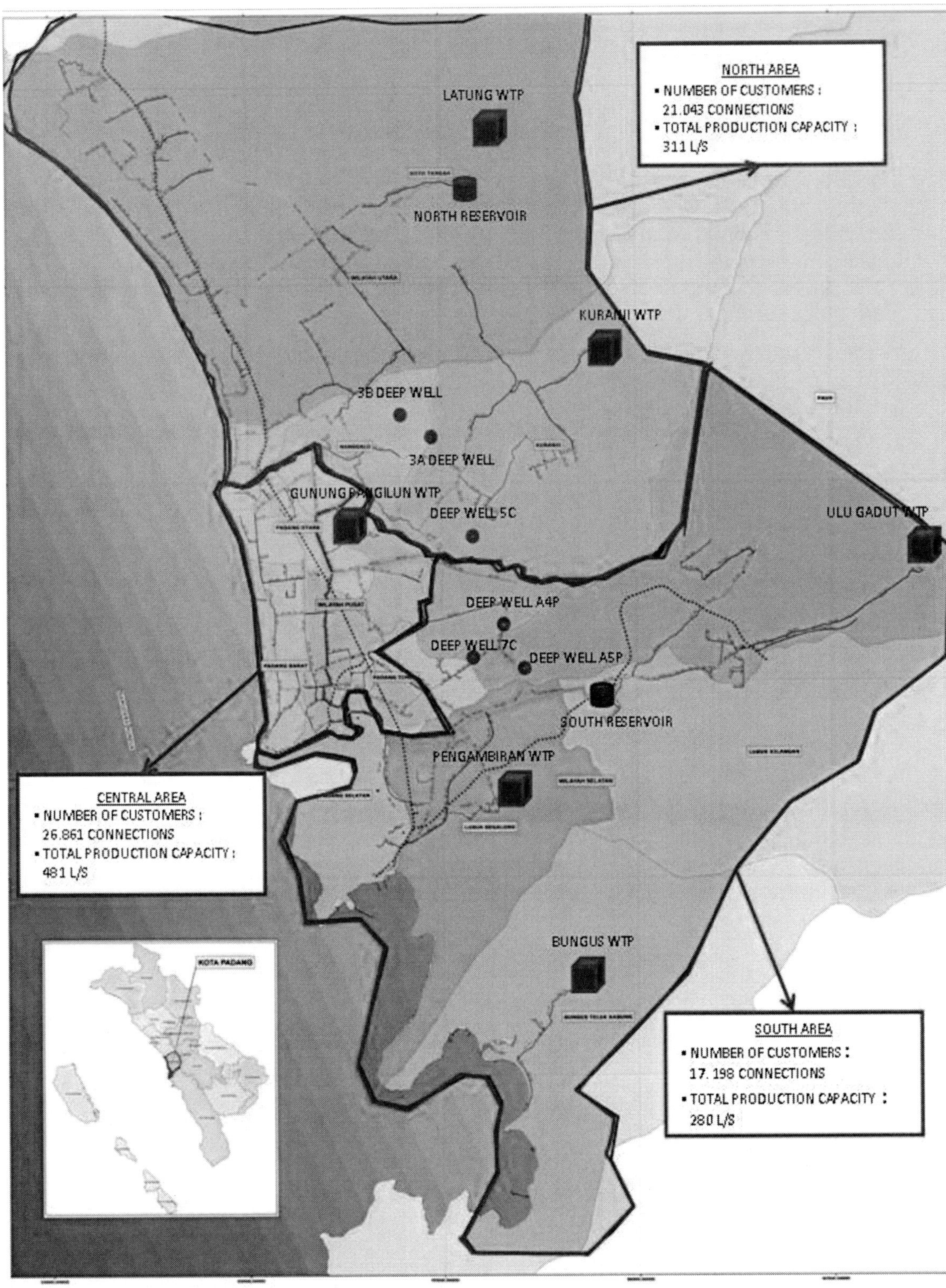

Figure 5.1. Padang Water System (Courtesy of Padang Water Department)

Table 5.1. Padang Water Supply

No.	Sources	Location	Production Capacity						Reservoir (m³)	Transmission System
			Installed Capacity			Effective Capacity				
			WTP (l/s)	Deep Well (l/s)	Total (l/s)	WTP (l/s)	Deep Well (l/s)	Total (l/s)		
(a)	(b)	(c)	(d)	(e)	(f)=(d)+(e)	(g)	(h)	(i)=(g)+(h)	(j)	(k)
I.	**CENTRAL AREA**									
1.	Gunung Pangilun WTP	Gunung Pangilun	500	-	500	481	-	481	-	pumping
2.	Gunung Pangilun Reservoir	Gunung Pangilun	-	-	-	-	-	-	3,000	gravitation
	Sub - Total (I)		**500**	**-**	**500**	**481**	**-**	**481**	**3,000**	
II.	**NORTH AREA**									
1.	Latung WTP	Lubuk Minturun	290	-	290	249	-	249	-	pumping
2.	Kuranji WTP	Guo Kuranji	40	-	40	38	-	38	-	gravitation
3.	3A Deep Well	Sungai Sapih	-	25	25	-	12	12	-	pumping
4.	3B Deep Well	Sungai Sapih	-	25	25	-	12	12	-	pumping
5.	North Reservoir	Lubuk Minturun	-	-	-	-	-	-	3,000	gravitation
6.	Kuranji Reservoir	Guo Kuranji	-	-	-	-	-	-	88	gravitation
	Sub - Total (II)		**330**	**50**	**380**	**287**	**24**	**311**	**3,088**	
III.	**SOUTH AREA**									
1.	Pegambiran WTP	Pegambiran	5	-	5	5	-	5	-	gravitation
2.	Ulu Gadut IA and IB WTP	Ulu Gadut	80	-	80	75	-	75	-	gravitation
3.	Ulu Gadut IIA and IIB WTP	Ulu Gadut	140	-	140	134	-	134	-	gravitation
4.	Jawa Gadut WTP	Jawa Gadut	20	-	20	19	-	19	-	gravitation
5.	Bungus WTP	Kp. Pinang Bungus	40	-	40	10	-	10	-	gravitation
6.	5C Deep Well	Kp. Kalawi	-	25	25	-	7	7	-	pumping
7.	7C Deep Well	Kp. Pisang	-	25	25	-	11	11	-	pumping
8.	A.4.P Deep Well	Kp. Pisang	-	8	8	-	3	3	-	pumping
9.	A.5.P Deep Well	Kp. Pisang	-	40	40	-	16	16	-	pumping
10.	Pegambiran Reservoir	Pegambiran	-	-	-	-	-	-	400	gravitation
11.	South Reservoir	Bandar Buat	-	-	-	-	-	-	3,000	gravitation
12.	Ulu Gadut Reservoir	Ulu Gadut	-	-	-	-	-	-	80	gravitation
	Sub - Total (III)		**285**	**98**	**383**	**243**	**37**	**280**	**3,480**	
	TOTAL (I+II+III)		**1,115**	**148**	**1,263**	**1,011**	**61**	**1,072**	**9,568**	

A steel pipe colocated with the Kurao Bridge along Jalan By Pass was damaged and was welded back together a few days after the earthquake (Fig. 5.13). Additionally, there were numerous broken service laterals, and many were not repaired during our visit. Desalination equipment was temporarily installed along the coast by the Australian Civil Services to provide potable water to the community (Fig. 5.14).

Mobile water purification units were sent out to communities without water during the month following the earthquake. Water was pumped directly from the river nearby and processed through the unit (Fig. 5.15).

Figure 5.2. A distribution line under a bridge that suffered little damage

Figure 5.3. Distribution lines suffered displacements at connection points on either side of bridges

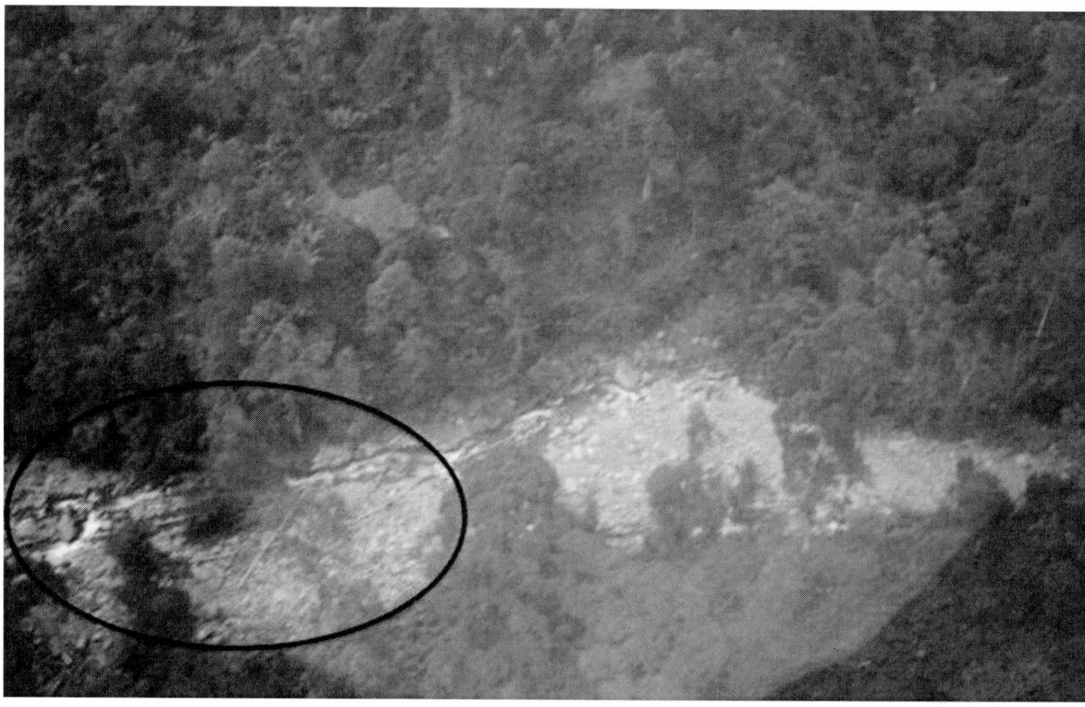

Figure 5.4. Water supply lines (circled area) in the southern area after the landslide

Figure 5.5. A ground view of the pipes shown in Figure 5.4

Figure 5.6. Supply material (PVC pipe) was flown in from Jakarta to meet the demand

Figure 5.7. Small leaks showed up during TCLEE visit.

Figure 5.8. Repaired buried steel pipe

Figure 5.9. Repaired buried steel pipe with a lateral

Figure 5.10. Repaired intake pipe to the treatment plant (Courtesy of Padang Water Department)

Figure 5.11. Damaged pipes being repaired at the Sumur Bor-3B water treatment plant (Courtesy of Padang Water Department)

Figure 5.12. Damaged pump at the Sumur Bor-3B water treatment plant (Courtesy of Padang Water Department)

Figure 5.13. A colocated broken pipe was being repaired. (Courtesy of Padang Water Department)

Figure 5.14. Desalination unit provided by Australian Civil Defence (Courtesy of Padang Water Department)

Figure 5.15. A mobile water purification unit was used to provide potable water to the communities without water supply. (Courtesy of Padang Water Department)

Record of Major Pipe Breaks

The following information and photos were provided by the Padang Water Department. This is not a complete record of the all the breaks in pipes; however, the information showed that more steel pipe than PVC pipe was damaged. The location of the pipe break was provided by the Padang Water Department.

5.3 Emergency Response

The water department building was severely damaged (Fig. 5.25) and had to be evacuated. Temporary administrative offices were set up in the one-story building behind the main building (Fig. 5.26). Service department desks were set up in the open area between the main building and the temporary administration building (Fig. 5.27). A temporary customer service desk (Fig. 5.28) was set up in front of the main building to handle customer inquiries and payments. The water department has its own radio system.

Dpn Minang Plaza Bawah Jembatan

Figure 5.16. Steel pipe broken at the T-connection (Courtesy of Padang Water Department)

Jalan By Pass Dpn SMP 24

Figure 5.17. Repair to a steel pipe above ground (Courtesy of Padang Water Department)

Jalan Cendrawasih Dekat Lembatan

Figure 5.18. Repair to a steel pipe break (Courtesy of Padang Water Department)

Jalan Lon Latang LB Minyurun

Figure 5.19. Repaired PVC pipe (Courtesy of Padang Water Department)

Jalan Maransi Raya Simp Mushala Daril Jamil

Figure 5.20. Repaired PVC pipe (Courtesy of Padang Water Department)

Jalan Raya Ikur Koto Dpn #28

Figure 5.21. Repaired PVC pipe (Courtesy of Padang Water Department)

Jalan Yaya Siteba Dkt Jembatan Bik Kantor KAN

Figure 5.22. Repaired PVC pipe (Courtesy of Padang Water Department)

Seb Palinggam Pdg Selatan Mushaia Nurui Sabah 1

Figure 5.23. Repaired steel pipe damaged at elbow above ground (Courtesy of Padang Water Department)

Seb Palinggam Pdg Selatan Mushaia Nurui Sabah 2

Figure 5.24. Damaged PVC pipe (Courtesy of Padang Water Department)

Figure 5.25. The front of the Padang Water Department; note the damage to the column on the right.

Figure 5.26. Temporary office building for administration

Figure 5.27. Temporary service desks for dispatch and repair services

Figure 5.28. Temporary customer service desk set up in front of the main building; note the damaged column at the back of the service desk.

Most Padang Water Department employees responded immediately to their office. Two days after the earthquake, there was a meeting about how to deal with the impact. Additional employees were flown from the Jakarta Water Department to supplement the efforts. Staff was divided into two groups: evaluation and repair.

The Padang Water Department also borrowed 33 water trucks from Jakarta and water departments from other cities. Local government provided cell phones for all managers and some staff, and hand-held radio units provided ample coverage after the earthquake between dispatch services and repair teams.

There was minor damage to one plant and no damage to reservoirs; however, the power outage interrupted pump operations at wells and plants. The water department was able to resume service by using the two in-house generators and four generators provided by UNICEF.

More than 1900 leaks were reported throughout the Padang area in the month following the earthquake. By the end of October 2009, 60 percent of the households had running water.

5.4 Recommendations

It is estimated that about 68 percent of the damage occurred where households connect to the network, and 22 percent of the damage observed was in PVC connections.

Following are some recommendations for consideration:

1. Restrain joints, such as gasket joints, to help reduce leaks and prevent pipes from separating. Welded joints are recommended for steel pipes.
2. Have backup power and fuel storage at all treatment plants, pump stations, and wells to keep the system operational.
3. Increase storage at multiple locations so that supplies are readily available during emergencies even if some supply facilities are damaged.
4. Keep stocks of repair material available for a quicker response.

5.5 Acknowledgments

The authors appreciate the support of the Padang Water Department. Its staff was very helpful and reviewed the details of their recovery with us, providing us with maps and photos of their effort. Mr. Ir. H. Azhar Latif, head of Padang Water Department, took time out of his busy schedule to ride with us in the helicopter and showed us the location of the transmission pipe damage and the treatment plant. His operations manager, Mr. Harry Satria, provided us with the water system information for the whole city. These two gentlemen are the keys of the success for this chapter.

Unless otherwise specified, all photos are taken by the ASCE/TCLEE team members.

Chapter 6
TELECOMMUNICATION SYSTEMS

From the beginning of this century, the telecommunication system in Indonesia has been undergoing fast-paced changes. Multiple service providers were allowed to compete in this market resulting in multiple services offerings at lower costs.

Although Telkom is still Indonesia's largest service provider, its market share is slowly being eroded by new competitors entering the market. The competitors are Indosat, Bakrie Telecom (Esia), Batam Bintan Telekomuniasi, Mobile-8 Telecom, Ceria, Smart, and XL Axiata, to name a few. Only four of these provide fixed-line (landline) phone services—Telkom, Indosat, Bakrie Telecom, and Batam Bintan Telekomunikasi. These four also provide all other telecommunication services including data, Internet, fixed wireless, and cellular phone service (commonly called mobile phone service in Indonesia). The remaining companies provide mobile phone and Internet services.

All Indonesian islands are interconnected via submarine cables and microwave. Recently, satellite connections have also come into use.

Cellular phone service is the fastest expanding market in Indonesia. The penetration rate is very high. Due to geographical nature of the country and dispersed population, fixed wireless is also quite popular. Therefore, there is a buildup of base transceiver stations (BTS)/cell sites around the country, especially on the biggest island, Java.

6.1 Description of System

The network configuration of the telecommunication system in Indonesia is basically the same as found elsewhere in the world. Most of the telecommunication equipment is imported from Europe, the United States, Japan, and recently from China. The basic elements of network components can be found in all BTSs and central offices (COs), which are also called strategic technology offices (STOs) in Indonesia.

Fiber optic cables are widely implemented to meet the demand for speed and bandwidth for Internet services. Optical fiber cable ring configuration is also deployed to provide a more efficient and reliable operation. Most hotels have Wi-Fi capability to meet the Internet services need of travelers.

The major service providers, such as Telkom and Indosat, have established different brand names for their various services. For example, Telkomsel is the brand name for Telkom's mobile phone service, and Telkom Flexi is the brand name for its fixed wireless service.

Colocating of service providers at point of presence (POP) is enforced by law, similar to North America.

In major cities such as Jakarta, BTSs mobile phone sites are sometimes installed on roofs of tall buildings to gain coverage. Similar to those in North America, the battery power reserve for these sites are usually low; that is, they have a short duration power supply after a power outage. In most cases, the battery backup is about 3 hours for BTS and about 6 hours for STO. Backup power generators are also installed in STOs to provide power to the equipment during a long duration power outage. Most STOs have BTSs colocated at the same site and share the building housing mobile phone equipment.

Standalone steel lattice towers are common for BTSs in rural areas.

6.2 Overview of System Performance

Fixed-line phone systems (including fixed wireless phone systems) and mobile phone systems did not perform well. Despite the widespread damage to the telecommunication networks, the recovery was reasonably fast due to the quick action and long hours of hard work of the service crews. In addition, external support from Jakarta to bring in the needed resources helped to speed up the process. Most services recovered within 3 days of the earthquake. The interruption of telecommunication service in the first two days after the earthquake created undue chaos within the emergency response services.

Telkom's offices and three STOs experienced severe damage. Many wiring circuits inside the STOs were damaged causing longer process to activate the generators to power the equipment. Most of the main switching centers (MSCs) were in good condition one day after the earthquake. Due to structural damage to the administrative offices in Padang, temporary offices were set up in an indoor tennis hall.

Many temporary BTSs were set up within Padang to provide mobile phone service when damaged BTSs could not be repaired. This type of unit is called cell on wheels (COW). It is the fastest means of replacing a heavily damaged BTS to restore service. However, the capacity of a COW is low, so call congestions and dropped calls occurred often in the areas covered by COWs.

Failure modes observed in STOs are

- damaged batteries,
- damaged building, and
- power outage.

While the failure modes of the mobile phone system are

- collapsed building with BTS installed,
- power outage,
- damaged power equipment, and
- generator problem.

The distribution system failure due to collapsed buildings on aerial cabling was quite significant.

One mobile phone service provider had 10 BTS antennas on rooftops of collapsed buildings damaged, and 28 BTSs had various power related problems. While another mobile phone service provider had so much damage to the mobile network that it required 2 days to restore service to normal.

The day after the earthquake, 42 out of 90 BTSs in one service provider's fixed wireless system in West Sumatera were out of service due to a prolonged power outage. In Padang, 47 of the 90 BTSs were not functional due to various problems. Most of the affected BTSs were back to normal 2 days after the earthquake.

6.3 Wireless System

In addition to observations by the ASCE/TCLEE team, this section includes information supplied by the fire department and a cellular service provider in Padang.

Cellular Base Stations (BTS)

Many BTSs have power supply units installed outside their buildings, which are air conditioned for the radio and controller units. A few BTSs were deployed with a self-contained, environmentally controlled unit within. Figure 6.1 shows one of these installations. Note the cabinet wall thickness; it is designed to reduce heat loss or gain. The cover protects service staff from the elements. Although these cabinets are designed with locks, a base perimeter wall provides added security. The majority of the BTSs are unmanned; that is, there is no attendant at the base. Monitoring is remotely done via data connection to a central control center within a region.

In addition to electric power outages, toppled equipment caused BTSs to go out of service. When equipment is not anchored or is inadequately anchored to resist seismic load, toppling is most common failure mode. The equipment is also installed on a structural platform about 0.7 m above ground, which amplifies the seismic load on the equipment. The platform is designed to provide vertical load capacity; there is no bracing to provide rigidity to resist lateral load (Fig. 6.4). The platform is a raised floor providing cable access to the bottom of the cabinets (Fig. 6.6). It was noted that heavy power cabinets with moveable components inside toppled. Figure 6.2 shows the failure of two unanchored power cabinets. Figure 6.3 shows the unsecured batteries inside, which might have caused the cabinet to tip over. A power cabinet anchored on a concrete pad in a BTS site had about 8 to 9 cm uplift but did not topple (Fig. 6.5). It appears that the anchor was inadequate to resist uplifting.

As Indonesia is frequented by monsoons with high wind speed, the towers are all designed to withstand a big wind load. Therefore, no standalone towers collapsed; however, BTSs on commercial buildings did.

Figure 6.1. BTS with a self-contained unit, no building is required. The base is still enclosed with a wall for security. (Courtesy of Thomas Heriyanto)

Figure 6.2. Toppled power cabinets at a BTS site outside of Padang. (Courtesy of Thomas Heriyanto)

Figure 6.3. Damage to the batteries in the cabinet close to the cracked wall in Fig. 6.2. (Courtesy of Thomas Heriyanto)

Figure 6.4. One of the two cabinets toppled at this BTS (Courtesy of Thomas Heriyanto)

Figure 6.5. The power cabinet at this BTS site is anchored on a concrete pad. (Courtesy of Thomas Heriyanto)

Figure 6.6. This toppled cabinet was saved by the wall. Note that there is no anchor for the equipment cabinet. (Courtesy of Thomas Heriyanto)

Antenna Towers

It was observed that the steel lattice style standalone towers performed well (Fig. 6.7). Wind load is the primary design factor. However, towers on roofs of commercial buildings shared the fates of the buildings. In most cases due to spectral amplification, the antennas mounting structures were damaged.

Figure 6.7. Typical standalone steel lattice tower shows good performance

Figure 6.8. Roof-top mounted BTS antenna poles failed when the building failed. There were three service providers located on the roof of this 5-star hotel in Padang.

Figure 6.9. When the second story of this building collapsed, the tower was damage and power was cut off from this site. (Courtesy of Thomas Heriyanto)

Figure 6.10. The BTS tower on the roof (left side) of this building that collapsed (Courtesy of Thomas Heriyanto)

Figure 6.11. Detail view of the collapsed tower and antennas (Courtesy of Thomas Heriyanto)

Figure 6.12. The tower and the equipment housing do not look damaged, but power was cut off when the building collapsed. (Courtesy of Thomas Heriyanto)

Figure 6.13. This BTS tower collapsed when the buildings around it collapsed in the earthquake. (Courtesy of Fire Department, Padang)

Figure 6.14. Due to strong shaking, one of the towers on the roof of this building was damaged. The building had no structural damage. (Courtesy of Thomas Heriyanto)

Figure 6.15. The BTS tower was undamaged, but with the building partially collapsed, this BTS site was out of service.

Figure 6.16. The BTS tower was damaged when this building collapsed. (Courtesy of Fire Department, Padang)

Backup power

For all remote offices and BTSs other than those colocated with central offices or main switching centers, battery reserve power is the only means of powering equipment during a commercial power outage. The reserve power lasted about 3 to 6 hours. Power generators are not deployed to remote offices or BTSs.

Indonesia is a tropical country with high ambient temperature, and cooling the electronic equipment adds to the electric power consumption. Equipment had to be shut down at a couple of sites due to overheating resulting from a cooling failure.

To cope with both power outage and power equipment failures, the service providers had to bring in power generators to maintain system operation. Head offices of all the service providers in Jakarta had to send mobile power generators to Padang to restore service. The emergency response process for these companies was efficient and within one day supplies were available at sites. With the widespread damage, repair crews worked long hours and slept onsite to reduce travel time to recover service quickly.

Figure 6.17. Generator were set on small flatbed trucks to be deployed.
(Courtesy of Thomas Heriyanto)

Figure 6.18. Manual unloading of generator set from truck.
(Courtesy of Thomas Heriyanto)

Figure 6.19. Technicians set up the generator and connect it to the power equipment onsite. (Courtesy of Thomas Heriyanto)

Temporary BTS

Many BTS sites installed on roofs of commercial buildings were out of service due to extensive building and equipment damage. To maintain coverage in these areas, mobile BTSs were deployed. Some these units were shipped in from Jakarta. The units setup relatively quick, but the capacity of each unit is much smaller than the original BTS it temporarily replaced. However, this is the fastest means of recovering telecommunication services during an emergency.

These units are self-contained with all the necessary equipment, a power generator, and tower structure to establish a functional BTS.

Figure 6.20. A mobile BTS cell on wheels (COW) in a container. Note the extended footings providing a structural foundation to support the tower.

Figure 6.21. The antenna tower of the COW in Fig. 6.20 is a steel telescopic cable guided tower.

Figure 6.22. Another type of COW with the power generator outside the unit (Courtesy of Thomas Heriyanto)

Figure 6.23. A unit similar to the one in Figure 6.20.

6.4 Telecommunication Building and Central Office

Telkom administrative offices in Padang sustained extensive structural damage but no casualties. The buildings, often referred to as the twin buildings as they have identical external features, are about 100 m apart. The north building is on the right in Figure 6.24.

Three STOs had damage to their power and cabling system. All were back to operation within 2 days of the earthquake.

Administration Office

The staff set up offices and customer service desks in tents outside the administration office building. The indoor tennis court between the buildings was used as a temporary operation center (Fig. 6.24).

The TCLEE team visited one of the damaged administration office buildings. There were significant structural damage and most of the nonstructural components inside were destroyed. The two buildings had nearly identical damage patterns on the outside (Fig. 6.25 and 6.26).

Figure 6.24. Twin Telkom buildings

Figure 6.25. The ground floor of the north building has a higher ceiling height than the upper floors.

Figure 6.26. Note the identical crack lines on the exterior wall of the south building

Figure 6.27. The north building interior had extensive damage to both structural and nonstructural elements.

Figure 6.28. Customer service desks in tents outside the north building

Strategic Technology Offices (Central Offices or Main Switching Centers)

Inconsistent installation practices were observed in the equipment offices visited, particularly in the equipment anchoring. Equipment that was part of the same network component was anchored in one room but in not another. This may be a result of different equipment providers having different installation practices.

Batteries were not secured on the racks to prevent them from strong shaking damage, and some of these racks were not anchored. The clean battery room indicated that the batteries were damaged, and the floor did not show any anchors were used (Fig. 6.29). Additionally, batteries were sharing the same space with telecommunication service processing equipment. In North America this is not allowed. Figure 6.30 shows the discarded battery racks.

Unsecured equipment placed too close together can result in collateral damage. Figure 6.31 shows the door of a piece of equipment that was damaged by a falling battery. The door most likely prevented major damage to the electronics, however. The batteries were not secured on the rack and the rack was not anchored (Fig. 6.32).

Many circuit packs were damaged in this earthquake (Fig. 6.33).

Due to extensive damage to the raised floor and the cabling within one of the equipment rooms where equipment was not anchored, the equipment had to be relocated (Fig. 6.34).

Figure 6.29. Equipment room where lead acid batteries had been damaged and removed. The battery racks that were installed on the raised pads were not anchored.

Figure 6.30. The battery racks were discarded just outside the battery room.

Figure 6.31. Replacement batteries were placed on the floor. The fallen batteries sliced the perforated equipment frame door surface.

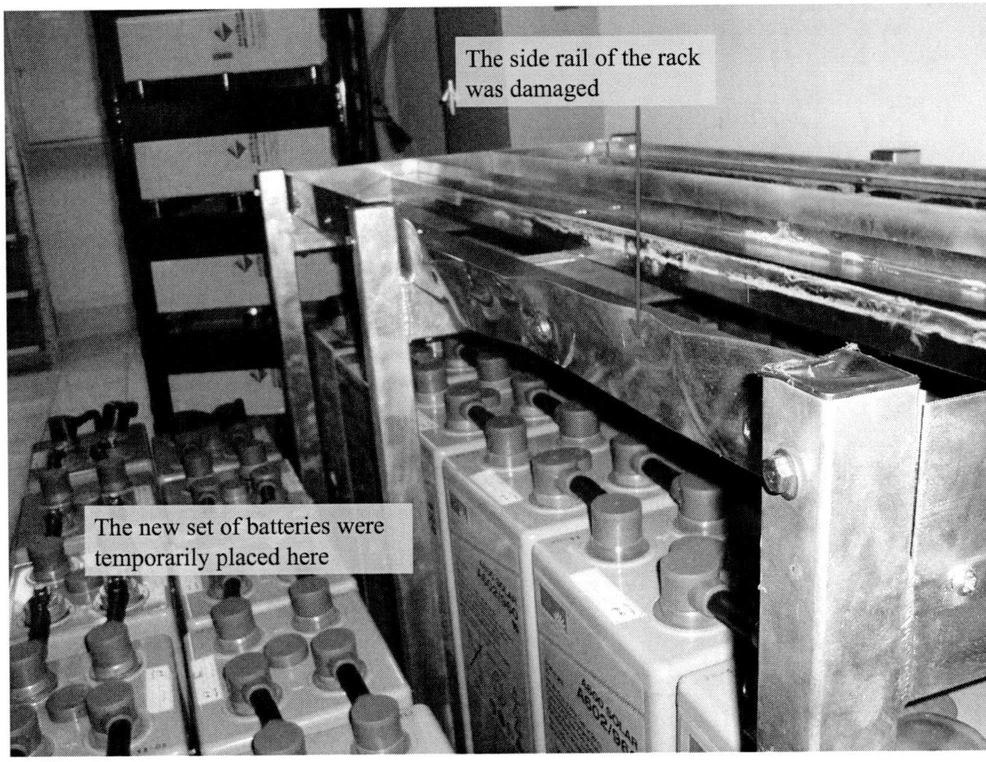

Figure 6.32. There was no provision for securing the batteries on the shelf. The bottom row of batteries was not constrained.

Figure 6.33. Damaged circuit packs and monitors were placed outside the equipment room.

Figure 6.34. Telecommunication equipment in this room was being relocated due to extensive damage to the equipment, cabling, and raised floor.

Figure 6.35. Equipment on this raised floor was secured to the concrete floor using a U-channel metal bracket with adjustable height under the raised floor.

Figure 6.36. Unanchored battery rack

The same equipment building housed equipment that was anchored to prevent toppling failure. Figure 6.35 shows equipment on a raised floor that was anchored via a metal bracket under the floor.

Different battery installation practices were observed. Figures 6.36 shows one installation in a small space with other equipment. This battery rack was not anchored. Overturning resistance was basically achieved by weight of the batteries on the bottom row; however, during the strong shaking of an earthquake, these racks could move around and impact objects close by causing equipment failure.

Figure 6.37 shows another battery installation in the same building as the one shown in Figure 6.36. These batteries were also unanchored and were configured in tall units that might overturn during an earthquake.

Figure 6.38 shows a new battery bank installation with anchor provisions. The anchors seem undersized, and the batteries are not separated by resilient material to prevent impact damage during shaking. However, this is better than unanchored equipment installed close to other equipment.

Figure 6.37. Batteries resting on a metal frame base with rubber feet and tied together via the buss bar connections between batteries

Figure 6.38 An installation that shows promising progress in preventing loss

6.5 Transmission and Distribution System Damage

The transmission network for fixed-line phone services had no reports of any significant damage. The distribution system damage was concentrated mainly around collapsed buildings. Although the collapsed buildings did not need phone service, the severed cables might include some transmission cables that would compromise the transmission network integrity.

6.6 Major Observations and Recommendations

The observations and conclusions after visiting a small sample of telecommunication equipment installations will not accurately represent all installations in Indonesia. The intention of this section is to provide some basic good engineering practices for seismic installations to reduce both direct and indirect losses.

Implementing good seismic installation practices for all network components can ensure loss reduction to both equipment and service.

The following list details proven good seismic installation practices:

1. Provide adequate anchoring of all network equipment to prevent overturning.
2. Secure all equipment on shelves to prevent falling off the shelves.
3. Bolt adjacent equipment frames together to prevent impacting.
4. Allow adequate cable slacks exiting the equipment frame to overhead cable racks to prevent cable and connector damage.
5. Secure batteries to the rack, and space them with a piece of Styrofoam.
6. Adequately anchor battery racks.
7. Consider the spectral amplification caused by the raised floor when anchoring equipment on a raised floor.
8. Separate battery and power equipment from the telecommunication processing equipment in different rooms.
9. Develop a standard on seismic installation to provide consistent performance in an earthquake.
10. Populate telecommunication equipment frames from the bottom to the top, so that the frames are not top heavy. Top-heavy equipment will raise the center of gravity of the unit that creates a larger overturning moment.
11. Re-evaluate all BTS installations on commercial building rooftops to improve the survivability of the mobile phone network.

The following list details how to improve BTS sites performance on commercial building rooftops:

1. Deploy more microcells that connect back to the BTS rather than the commercial building roof.
2. Evaluate structural design of the commercial building before deploying BTS.
3. For existing sites, install quick connect capability to the MSC when a COW is needed after a disaster.
4. Increase the battery backup to allow the site to operate longer without commercial power so that small mobile generator can be brought in before the site stops operating.

The coupling effect of lifelines interdependence is extremely important in planning and in upgrading the telecommunication networks. Electric power again proved to be one of the critical factors in providing a reliable and resilient telecommunication network.

On interdependence, the recommendations are:

1. Develop a business plan to install backup power generator in all sites.

2. Develop a procedure to test and dry run a power outage exercise annually.
3. Establish strategic generator fuel storage locations for quick access to fuel supply.
4. Store COWs at these locations as well.
5. Establish alternate route access to remote sites.
6. Establish emergency services procedures to assist both essential services (such as police, fire fighters, and medical services) and public services in the disaster area.

A well-balanced risk management program to reduce losses in both services to the customers and the capital assets will ensure business continuity after a disaster such as an earthquake. The above recommendations are based on the team's brief site visits. A detail walk down evaluation of each site will provide a more complete list of vulnerable items to develop a plan to mitigate damage.

6.7 Acknowledgements

I am indebted to many individuals who assisted us in this investigation. In particular, Prof. Oey's (Hong-Sioe) effort setting up local contacts for the team prior to arriving Padang is much appreciated. I am extremely grateful to Mr. Roni Handawi, who works in Jakarta and accompanied us to Padang organizing meetings with the telecom service providers so we could collect performance information. The quality time provided by Mr. Syahril and Mr. Thomas Heriyanto to talk to us about the performance of their networks is much appreciated. This chapter would not be useful without Mr. Thomas Heriyanto's photos and permission to use them.

Chapter 7
SEAPORTS AND AIRPORTS

Sumatra is the biggest island of the Republic of Indonesia, which comprises of more than 17,000 islands. Seaports are one of the major means of transporting products internally and internationally. There is always a seaport close to a populated area.

Teluk Bayur Port is the largest and busiest port on Sumatra's west coast. It is located just south of Padang. This port handles most of the import and export shipments.

Airports are developed when air travel becomes more popular. There are many domestic interisland flights. Most international flight to Indonesia come through Jakarta International Airport, which is on the island of Java.

Padang, located in the West Sumatra Province, is a major city on the island of Sumatra, and it has its own international airport called Minangkaban International Airport with an international airport code of PDG. This airport located in Ketaping north of Padang was opened in July 2005, replacing the Tabing Airport. This airport has international flights departing and arriving from cities in Southeast Asia. The airport is managed and operated by the Indonesian Department of Transport, which is also responsible for the air traffic services in Indonesia.

7.1 Teluk Bayur Port, Padang

Teluk Bayur Port (Figure 7.1) is located 5 km south of the city of Padang on Bayur Bay. It is formerly known as Emmahaven Port. Shipments are distributed by both road and railway to their respective destinations within the island of Sumatra. The port is owned and operated by Indonesian Port Corporation II, a government-owned company.

The port has three piers. One is for conveyor loading and unloading of materials. The other two service container ships. Fishing boats moor at the fishing village on the east side of the island. Since the port is located in a bay, it would avoid direct hits by a tsunami from the west.

The port also has a large tank farm for petroleum products (Fig. 7.2).

The only observed damage in the proximity of the port is the new administrative building parking lot. The edge of the pavement showed a surface crack caused by lateral spreading (Fig. 7.3).

Figure 7.1. Teluk Bayur Port (Photo courtesy of Yusuf Ederis)

Figure 7.2. Tank farm at the Teluk Bayur Port (Photo courtesy of Yusuf Ederis)

Figure 7.3. Lateral spreading damage to the parking lot and a minor soil slumping at this new administrative building (Photo courtesy of Yusuf Ederis)

7.2 Minangkaban International Airport, Padang

The airport terminal building sustained minor damage to the roof, which resulted in closing the airport for 12 hours after the earthquake. It was open the morning of October 1. There was nonstructural damage inside the terminal, such as fallen ceiling tiles. The traditional bullhorn style of A-frame roof had minor damage. The air traffic control tower was not damaged.

The terminal building and the control tower are reinforced concrete frame structures.

Figure 7.4. Minangkaban International Airport

7.3 Acknowledgements

The author appreciates the generosity of the Pandang Mayor Dr. H. Fauzi Bahar, who provided us with a helicopter to survey the earthquake impacted locations.

Chapter 8
EMERGENCY PREPAREDNESS

In all provinces in Indonesia emergency response follows the national guidelines. There are also several grass root organizations that responded to the earthquake. This chapter provides an overview of each emergency response organization and its goals and objectives. Response to this earthquake was the first test of effectiveness for some of the organizations and provided them an opportunity to review their response and strategize their efforts.

8.1 National Agency for Disaster Management

Indonesia has an established program for disaster response that is initiated by the Badan Nasional Penanggulangan Bencana (BNPB) or the National Agency for Disaster Management.

National Agency for Disaster Management	**BNPB** Badan Nasional Penanggulangan Bencana
Provincial Agency for Disaster Management	**BPBD Propinsi** Badan Penanggulangan Bencana Daerah
District/Municipal Agency for Disaster Management	**BPBD Kabupaten/Kotamadya** Badan Penanggulan Bencana Kabupaten/ Kotamadya

Indonesia's national agency has issued public policy and guidelines that include measures for preparedness, mitigation, and recovery. The national agency provided training for the local government after the guidelines were updated in 2006.

Provisional agencies can modify the national guidelines to meet unique local circumstances but still need to be in general compliance with the national plan. Governors have to pass a resolution to adopt the BPBD guidelines for their provinces. The mayor is in charge of reporting the level of damage to the national authorities. The national agency responds to the mayor, who is in charge of the operations of the Posko Utama (Disaster Management Center).

Disaster Management Center in Padang

BPBD was launched by the Office of Mayor of Padang in February 2009, and consultants were hired to train department managers. BPBD had identified the need to hire 50 to 60 employees. However, as the agency was established in February 2009, only 35 employees were hired by September 30 of that year.

The BPBD building was equipped with tents, rubber boats, food (for the public), and a generator. The police, fire department, and BPBD have hand-held radios for communication. Lists of equipment for various departments are also available at this office.

The Posko Utama, which was launched for emergency response in the month of October 2009, was held in the mayor's office as the BPBD building was severely damaged during the earthquake.

After the initial response to the emergency is completed, Posko Utama will be dissolved, and the concentration shifts to rehabilitation and reconstruction under the BPBD general guidelines. Due to the extent of destructions in Padang, BPBD will be supported by the other municipal agencies in this mission. In Padang, the BPBD office has developed a general policy for rehabilitation and reconstruction of the city and emergency action plan.

Goals of Padang BPBD

The Pandang BPBD has identified eight goals:

1. Relocation of city government center to the new proposed location
2. Recovering the city economy and revitalizing of Pasar Raya (Grand Market) and satellite markets
3. Re-organization of the urban public transport network
4. Revitalizing the old city center areas
5. Rehabilitating education and health infrastructures and facilities
6. Recovering private residential areas and the settlement's infrastructure
7. Recovering the community's psychological aspects and implementing disaster preparedness
8. Reviewing and revising the city spatial (land use) planning (RTRW) and city mid-term development planning (RPJMD)

Response Overview

During the earthquake, six employees were in the office. They left the building immediately and called the national office to get information about the earthquake they had just felt and its magnitude. They connected their computers using onsite generators to find out if there was any chance of a tsunami.

This office received many calls immediately after the earthquake struck. Questions were mostly about the intensity of earthquake and potential for a tsunami. People were panicked about the potential for a tsunami and were trying to escape to higher grounds. This caused heavy congestion on all major streets. BDBP personnel tried to calm the people and send them back to their houses.

Employees in the office had a hard time reaching their families because of limited cell phone operation. After verifying that there were no threats of tsunami, they left to check on their families. After that they reported back to work contacting and coordinating with other departments—police, fire, hospitals, and national government.

Various departments provided updates on damage, fatalities, and response activities, and the Posko Utama recorded the information in each of 11 districts of Padang. Response was prioritized based on the number of victims in any given location. The total number of fatalities reported by this office was 383 in Padang and about 1100 in the West Sumatra area.

Posko Utama was the main coordination point for all the non-governmental organizations (NGO) and donations from other countries. It maintained a detailed inventory of cash and donation items received and distributed.

Recommendations

The author provides the following recommendations:

1. Focus on raising awareness and preparedness as a main goal.
2. Develop and enforce better construction standards.
3. Add more equipment (boats, boots, etc.) and more communication devices at all departments.
4. Posko Utama should be announced as the central receiving place for all NGO's rather than various offices.

Fig 8.1. Posko Utama (Emergency Management Center) set up in Padang

Fig. 8.2. Detailed information was posted to help prioritize the response.

Fig 8.3. Distribution of Emergency Supplies

8.2 Tsunami and Disaster Mitigation Research Center (TDMRC) of Syiah Kuala University, Banda Aceh, Indonesia

TDMRC was established following a decree signed by the Rector of Syiah Kuala University in October of 2006 in response to the 2004 earthquake and tsunami that devastated Aceh Province in northern Sumatra Island. The primary goal of TDMRC is to conduct applied studies in disaster risk reduction (DRR) that produce a strategic concept in establishing people's awareness and the building community's resilience during disasters with a special emphasis on earthquakes and tsunami.

To achieve this goal, TDMRC formulated seven strategic initiatives:

1. Encouraging national and international collaborative research and education in DRR
2. Compiling all disaster-related data for Aceh and Indian Ocean region to be organized in a virtual museum that can be accessed globally
3. Developing materials related to DRR for a community disaster education program
4. Conducting training for teachers and integrating DRR into school curriculum
5. Serving as the "Think Tank" for the Aceh government in implementing DRR programs and designing strategic plans for disaster management
6. Fostering a TDMR–community partnership in DRR
7. Developing the Aceh region as an international disaster research laboratory that leads to the establishment of TDMRC as the Center of Excellence for Indian Ocean studies

Mission: To be an open, prominent institution in multi-disciplinary applied research, education and training, data and information delivery and knowledge building, and professional services in the area of disaster management.

Vision: Protecting communities through effective science-based disaster risk management.

TDRMC implements its programs through four divisions: Research and Development; Knowledge Management; Education, Training, and Advocacy; and Professional Services.

Since 2006, TDMRC has organized an Annual International Workshop and Expo on Sumatra Tsunami Disaster and Recovery (AIWEST-DR) to document the lessons learned from the post-tsunami rehabilitation and reconstruction activities in Aceh and other tsunami-affected regional countries and to review how this rebuilding effort contributes to building disaster-resilient communities and better environments as well as preserving the legacy of the worst natural disaster of the century for future generations. This international workshop brings together experts, researchers, policy-makers, and practitioners from around the world sharing their knowledge and experiences that improve earthquake and tsunami disaster management strategies, thereby reducing risks in tsunami-prone regions. This workshop is organized in cooperation with the Research Center for Urban Safety and Security (RCUSS) of Kobe University, Japan.

Currently, TDMRC is linked to many national and international institutions working in DRR such as the Ministry of Research and Technology; Ministry of Foreign Affairs; Indonesian Institute of Science; Ministry of Home Affairs; National Development Planning Agency; Agency for Meteorology, Climatology, and Geophysics; Bandung Institute of Technology; United Nation–International Strategy for Disaster Reduction (UN-ISDR)/International Recovery Platform; American Society of Civil Engineers (ASCE)/Technical Council on Lifeline Earthquake Engineering (TCLEE); International Center for Water Hazard and Risk Management (ICHARM), Tsukuba, Japan; Pacific Tsunami Museum–University of Hawaii at Hilo; CEDIM, Karlsruhe University, Germany; Stockholm Environment Institute (SEI); Tsunami Research Center, University of New South Wales, Australia; UNESCO: Oregon State University; Kyoto University, Nagoya University, Tokyo University, Vassar College, New York; University of

Pittsburgh; Earthquake Megacities Initiatives (EMI); Sea Defense Corp. Deltares Consulting (The Netherlands); University of Waterloo, Canada; University Sains Malaysia; University Kebangsaan Malaysia; Disaster Management Center of Sri Lanka; National Disaster Management Agency of India; All India Disaster Management Institute; Disaster Management Agency of Maldives; Indonesian Red Cross; Jakarta Tsunami Information Center; American Red Cross, KOGAMI, Padang; and others.

Core TDMRC staff members include faculty members of Syiah Kuala University, Banda Aceh, coming from various academic backgrounds not related to disaster management. However, for the past two years, some members have received short-term supplemental training on disaster management from research centers in Japan, Malaysia, and The Netherlands.

To date, TDMRC has trained many local school staff and volunteers and has collaborated with the Indonesian Institute of Science in establishing three model schools (elementary, junior high, and senior high schools) for disaster preparedness in Banda Aceh. This research center has also held a training seminar for 28 professors joined in the Canada-Sri Lanka University Consortium for Post-Tsunami Recovery funded by the Canadian International Development Agency (CIDA).

In October 2009, TDMRC participated in a post-earthquake reconnaissance survey in Padang in cooperation with ASCE/TCLEE. The head of the TDMRC Professional Services Division, Teuku Alvisyahrin, Ph.D., led the TDMRC survey team consisting of Purwandhi, Wan Akmal, and Diky Juliandi, while facilitating communications with Padang officials for the ASCE/ TCLEE team consisting of Alex Tang, P.E., and Sudi Shoja, P.E.

In March of 2010, TDMRC hosted two important international training workshops: 1) Disaster Management with a focus on Coastal Vegetation Management for Tsunami Disaster Mitigation in collaboration with ICHARM, Japan, and 2) the second South-South Cooperation Community-Based Development Academy on Disaster Risk Reduction in cooperation with United Nation Development Program, Bangkok. The latter workshop will bring participants from 12 regional countries.

8.3 Collaboration of University Professors and Students

Structural engineering professors and students from the following local universities joined hands and formed a volunteer response group to evaluate the status of the residential building immediately after the earthquake:

Andalas University, Padang Institute of Technology, Bung Hatta University, Ekasakti University (State University of Padang), Putra Indonesia University, and Institute of Industrial Technology. Febrin Anas Ismail, Ph.D., of Andalas University supervised the group while Arighan Amin, a civil engineering graduate student with the Andalas University, led the volunteer effort.

The coalition between the universities was formed in 2007 following another earthquake in the region. The goal of this effort was to provide an evaluation of all the residential buildings for the owners so they could go back to their homes.

Each university has a list of volunteers that were notified using a telephone-tree system with 50 volunteer coordinators. Announcements were also made over media. The day after the earthquake, 720 volunteers reported at the Andalas University campus. Their efforts continued for 14 days. Then the number of volunteers reduced to 50 after inspections were completed.

Volunteers were accompanied by members of government organizations such as Public Works and carried two forms of decrees form the municipality and the village director to officiate their work.

There were no consultants or contractors among this group. The Padang Chapter of Society of Consulting Civil Engineers of Indonesia plan to perform a survey and evaluation of commercial and public buildings. The ASCE/TCLEE team did not verify this information independently.

Inspections

Guidelines were developed the day after the earthquake and given to all engineers to ensure consistent evaluations. Buildings were evaluated and tagged by color prints for heavy (beyond repair), medium, and light damages. Colored prints were attached to the front of every building. The outside and inside of homes were inspected as much as possible.

Volunteers reported to the PW subdistricts and were accompanied by PW personnel. This effort was minimally funded. Volunteers were just given writing tablets and colored tags. They did not even have hard hats.

Process

Volunteers were asked to locate the owners, most of whom were staying close to their homes, to explain the process and the result of their evaluations. If an owner could not be located, the volunteer left a message with neighbors.

Volunteers were given survey forms to fill out at each site and rate the damage to homes. They were asked to post their colored ratings in front of each building and take a picture using their personal cell phone.

All forms and pictures were then turned in to the coordinator at Andalas University.

Coordinators were responsible for compiling the collected information, completing web filing, and providing statistics.

Conclusion

This group of volunteers completed inspection of 160,000 residential buildings in 14 days. The result of the inspections revealed that 53 percent of homes were heavily damaged and needed to be demolished, 38 percent were rated medium, and only 9 percent lightly damaged. Reports will be given to the Padang BPBD, which will forward them to the provincial BPBD for further action. In addition, they are usually required to report to their respective university as part of the record for the programs.

Recommendations

1. Finalize the inspection guidelines based on lessons learned from these inspection efforts for future use.
2. Document the process used including any improvements needed.
3. Hold annual training sessions for current and new volunteers. Issue certificates and identification cards for volunteers who complete the training sessions.
4. Update the volunteer list annually.
5. Secure items needed by volunteers such as hardhats and writing tablets in designated storage areas that are easily accessible.
6. Initiate communication with other local groups such as BPBD Propinsi and KOGAMI.
7. Expand collaboration efforts to all local universities in the Padang area and those on the island of Sumatra.

8.4 Komunitas Siaga Tsunami (KOGAMI)

KOGAMI is a community based non-profit organization with the goal of providing disaster preparedness training and education in Padang area. Ms. Patra Rina Dewi, M.Sc., founded KOGAMI in July 2005 after volunteering with Surf Zone Relief Organization (SRO) from San Francisco, US. Ms. Patra Dewi has been fully trained in United States. Staff has been trained by organizations such as Red Cross and SRO.

The KOGAMI mission has three goals:

1. Build organization capacity in disaster management.
2. Develop an integrated system for disaster management based on professionalism.
3. Network to optimize work performance.

KOGAMI also developed two business plans that earned them financial donations from SRO and the United Nations. KOGAMI's efforts received considerable community resistance because the volunteers were considered messengers of bad news. To deal with this matter, staff worked on improving their communication and presentation skills, invited specialists to present factual information, and developed better printed media.

Currently this organization has 14 staff and 50 field trainers and 250 registered volunteers. KOGAMI is concentrating on programs for four audiences:

1. Government officials
2. Government agencies, such as public works, transportation, police, and fire departments, to familiarize them with the emergency response plan and standard operation procedures.
3. Community group training for residents and villages to develop disaster preparedness teams and provide first aid training through trusted people in the community who are equipped with hand-held radio units
4. Schools to prepare a disaster response teams at every school

KOGAMI also provides psychosocial services for the population in large.

Chapter 9
FOLLOW-UP ASSESSMENT REPORT

By Alex Tang, Curtis Edwards, and Teuku Alvisyahrin

The ASCE-TCLEE, follow-up on lifeline improvement and recovery assessment team consisted of the following members:

- Alex Tang (ASCE), Team Leader
- Curt Edwards (ASCE),
- Teuku Alvisyahrin (TDMRC, Indonesia)
- Sudi Shoja (ASCE)

ASCE's Executive Committee initially suggested this project when TCLEE requested funding to investigate lifeline performance after the damaging earthquake in 2009. TCLEE accepted the challenge and planned the project one year after the earthquake with an execution date of 2 years after the earthquake. This is the first TCLEE attempt to investigate and assess improvements and progress in recovery after a major earthquake. TCLEE appreciates support and funding from ASCE senior staff its Technical Activity Committee (TAC). The team feels that this follow-up investigation was a worthwhile exercise that produced a great deal of useful information to be shared with engineering communities.

The logistics plans by Sudi Shoja and support by Teuku Alvisyahrin, Ph.D., head of the Professional Services Division of TDMRC and a professor at Syiah Kuala University, contacting the local lifeline authorities were key to the success of the trip. After all her hard work Ms. Shoja was unable to join the team in Padang because of a family emergency.

TCLEE/ASCE has created a high level of trust with the lifeline communities in Padang through this follow-up exercise. The general comment from the local officials was "Every organization came to collect information and never returned; what do they do with the information?" This visit answered their question.

9.1 Summary

Although the forecast called for rain and thunderstorms, we had little or no rain for the entire 3 days were in Pandang. After our first meeting with the mayor's assistants we were able to fill all our open slots in our meetings schedule. This included meetings with many departments that we were unable to get a response from in advance. For example, we were fortunate to meet a member of the water department board of directors, and with one phone call, a meeting was set up. The local lifelines authorities were all busy, but they knew that meeting us was not a waste of their time.

The follow-up information and lessons learned were based on the following questions:

1. Has there been any change to policy and operation after the earthquake?

2. What is your experience in your recovery process?

3. Is the city back to normal yet?

The answers were quite consistent among the different lifeline operators we talked to. Some operators had progressed more than the others; the fundamental difference was the availability of funding. Government operated lifelines such as the water system, did not make any improvements other than fixing what had been damaged.

Emergency Services, a new department formed after the earthquake, equipped themselves with rescue equipment. It also organized community volunteers and conducted training to support them in case of future emergencies. However due to limited funds, they unable to acquire adequate communication equipment for the various communities. This makes it harder for them to respond to emergencies. One of their many concerns is the inadequate number of tsunami evacuation signs that direct the residence where to go in case of tsunami warning. We suggested that the head of the department solicit donations from equipment manufacturers.

Based on the meeting with one of the telecommunication service providers in Padang, the improvement in this sector was significant, correcting the problem experienced during the last earthquake. We are positive that when their program is completed they will have a more resilient system.

More than 200 new schools were built to replace the damaged/destroyed schools. Most all of the new schools were funded by private corporations and accordingly the designs varied. One design method used a steel frame structure, while another used a reinforced concrete frame with unreinforced infill masonry. The new schools are designated as vertical emergency shelters in the city's tsunami zone. They are all the same height – three stories with a helipad on the roof for emergency response. Based on our observations, they are designed and constructed to a much higher standard and should perform during future earthquakes.

9.2 Emergency Response

A Badan Penanggulangan Bencana Daerah (BPBD) is a new provincial, district, or city agency for disaster management formed after the 2009 earthquake. For the city of Padang, the head of the agency is Dedi Henidal, Ph.D. (Fig. 9.1). Due to limited funding, many essential functions for this new agency were only partially completed. The agency is still housed in a temporary lot with cargo containers as offices (Fig. 9.2).

Figure 9.1. Meeting at the City of Padang Agency for Disaster Management office. From left to right, Henky Mayaguezz (BPBD), Curt Edwards (TCLEE), Drs. Dedi Henidal (BPBD Chief), Alex Tang (TCLEE), and Teuku Alvisyahrin (TDMRC).

Figure 9.2. Temporary offices of Padang Emergency Response Department

With the support of Jakarta, the agency organized a mental health center called healing home for kids. This new center was essential for the recovery effort and worked quite well.

The government provided monetary support to owners of houses that were damaged in the 2009 earthquake. For heavily damaged houses, 15M Rup were provided for each house, while for moderately damaged houses, the funding was 10M Rup. These funds were not enough for full restoration of homes, but provided some much needed assistance

Reconstructing damaged roads is still on going with only seven of the 20 roads on the reconstruction list completed. Again this is due to funding from the central government.

BPBD organized groups of 20 to 30 volunteers from each village as the backbone for disaster reduction and emergency services for the community. BPBD estimates that about 2,080 people need to be trained. So far only about 500 have been trained.

The problem with these groups is the lack of telecommunication equipment to share information during a disaster. Some groups are using "handy talkies," which have very limited coverage. We suggested that Mr. Henidal to contact telecommunication equipment manufacturers to ask for donations for these special groups, so they can improve communication during a disaster.

The water supply during the post disaster period was insufficient. In the future, Emergency Response Department officials are planning to set up additional water storage to handle post-disaster water shortages. One of the main reasons for this shortage after the 2009 earthquake was the lack of water tanker trucks. We later learned that more tanker trucks are out of service than in service due to a lack of maintenance and care (Fig. 9.3). The unused tanks were all stored in an empty lot near the reservoir by the water treatment plant and are likely not repairable. If these were available in times of emergency, there should be an acceptable level of water supply to the communities.

As mentioned, there were not enough tsunami evacuation signs. The BPBD plan calls for 2,000 more signs for the city. Some will direct people to go to vertical shelters; others will direct people to go to higher ground. See Figures 9.4 and 9.5.

After our meeting, the Emergency Response Department officials showed us the storage area for the emergency response equipment. They were placed in an open space at the back of concrete frame building (Fig. 9.6).

Figure 9.3. Out of service water tanker trucks

Figure 9.4. Tsunami sign that directs people to the vertical shelter (SMA-1 is the name of a school that is within 200 m)

Figure 9.5. Tsunami sign for the evacuation route

Figure 9.6. Emergency response equipment storage

9.3 Water System

Due to the potential structural collapse danger of the water treatment plant, we did not visit the plant during our investigation in 2009. We were able to see the potential danger this time (Fig. 9.7 to 9.11). The room where backup generators and power equipment were housed almost collapsed. Steel beams and poles are now used, providing structural support to retain a shelter for the equipment.

Figure 9.7. Water treatment plant building external damages were not fixed.

Figure 9.8. Inside the building, the repair work shown was the steel frame, which provided some structural support. Note that the wall remains the same as it was after the earthquake 2 years ago.

Figure 9.9. Another view of the fixes to the operation room where critical equipment is housed.

Figure 9.10. The damaged batteries were replaced with new ones, but they were not secured.

Figure 9.11. The transformer was on rail and was not secured. It is allowed to move around; the only stop is the edge of the floor opening.

This plant is more than 50 years old and is located on a cut and filled section of the hillside, which caused ground settlement that impacted the buildings and the concrete tanks (Fig. 9.12 and 9.13). With the poorly designed concrete structures and the poor soils conditions, we felt that it might not be able to hold up another similar event. We told them that the best option was to build a new plant.

The Water Department headquarters building sustained extensive structural damage. It was not repairable and was demolished. Figures 9.14 and 9.15 show before and after the demolition respectively.

The raw water intake control structure at the river near the plant needs a little renovation; otherwise it is functional (Fig. 9.16).

The water department should also initiate a maintenance program to rebuild the tanker trucks (Fig. 9.3) or find funds to purchase new water tanker trucks for emergency response purposes.

Figure 9.12. This step shows how much the ground around the building has settled, about 8 in. (~ 20 cm).

Figure 9.13. Evidence of ground deformation around the foundation of the water tank. A slight tilt is noticed by the water line along the edge of the tank.

Figure 9.14. The building on the left was the Water Department's main building. This photo was taken from south facing north, four weeks after the 2009 earthquake.

Figure 9.15. The damaged Water Department main building was demolished and debris removed. This photo was taken from north facing south, September 2011.

Figure 9.16. Water intake gate

9.4 Telecommunication

With the assistance and persistence of Teuku (Alvis) Alvisyahrin, the head of Professional Services Division of TDMRC, we were able to meet with the network operation manager of Telkomsel, a major cellular service provider in Padang. Figure 17 is a group photo taken after our meeting at Jl Khatib Sulaiman, where Telkomsel's main hub of is located. Mochamad Basyumi, manager of Pandang's network operations, spent the better part of his day guiding us through the main operation building. He provided excellent details on the damage and showed us the retrofit and equipment replacement improvements. The improvements were impressive, showing us that the engineers in Indonesia have a good sense of seismic design.

Mochamad Basyumi gave a personal guided tour of the building after a short meeting. This facility is not fully functional yet. Employees at this facility were transferred to another location. See Figures 9.18 and 9.19 for a view of the back of the building an aerial view of the facility. This is reinforced concrete building with very high ceilings on the ground floor. This is almost equivalent to a soft story building, which resulted in significant non-structural damage such as fallen ceiling tiles, light fixtures, and such. Shear walls (Fig. 9.20) were added to all floors to stiffen the structure and avoid similar damage in a future earthquake. Figure 9.21 shows the same corner of the room before the shear wall was added and four weeks after the 2009 earthquake. Damage to the ceiling in this accounting office is evident.

Figure 9.17. Meeting at Telkomsel, from left to right: Curt Edwards, Alex Tang, Mochamad Basyumi, and Teuku Alvisyahrin

Figure 9.18. Back of Telkomsel building on Jl Khatib Sulaiman in Padang

Figure 9.19. Aerial view of the Telkomsel facility still under construction

Figure 9.20. Ground floor of the Telkomsel facility, where shear wall (left hand side) was added

Figure 9.21. The same room (Fig. 9.20) after the 2009 earthquake, note the large floor to ceiling window on the right.

The team also visited the equipment rooms, where telecommunication equipment and power equipment are installed. Due to network planning changes and some equipment damage, new equipment and power plants are being installed. With the new equipment, this facility will become one of the major hubs of Padang's cellular service. Equipment frames were anchored and the battery plants were secured via structural steel frame or shelf-style cabinets anchored to the concrete floor. Figures 9.22 to 9.27 show the equipment and power rooms, as well as the anchoring details. Some improvements to prevent damage to the batteries were pointed out to Mochamad Basyumi; Figure 9.28 shows one of the recommendations to improve the seismic performance of the battery installation. Plastic straps were recommended for the shelf mounted batteries. Since this equipment is installed on upper floors of the building, the seismic load will be higher.

The top two floors are currently being use for employee recreation purposes and material storage (Fig. 9.29).

We then used the newly constructed the emergency evacuation stairs (Fig. 9.30). We noticed that the emergency escape doors were all locked by an energized magnet (Fig. 9.31). We asked Mr. Basyumi to check if the door locks will be de-energized when power to the building was not available. It is common to have a power outage immediately following an earthquake.

Figure 9.22. Equipment room, most of the rooms were quite empty like this one. There was a plan to populate the rooms with telecommunication equipment to build up redundant capability for the network in Padang.

Figure 9.23. New network equipment being installed

Figure 9.24. A typical frame anchoring arrangement

Figure 9.25. Fiber optic and DSL equipment

Figure 9.26. The unsecured batteries on the floor were the old batteries, which will be removed. The new batteries were installed on shelves. We recommended they be strapped down to the shelf.

Figure 9.27. The battery bank on the left will be replaced. Some of the batteries were damaged by the 2009 earthquake.

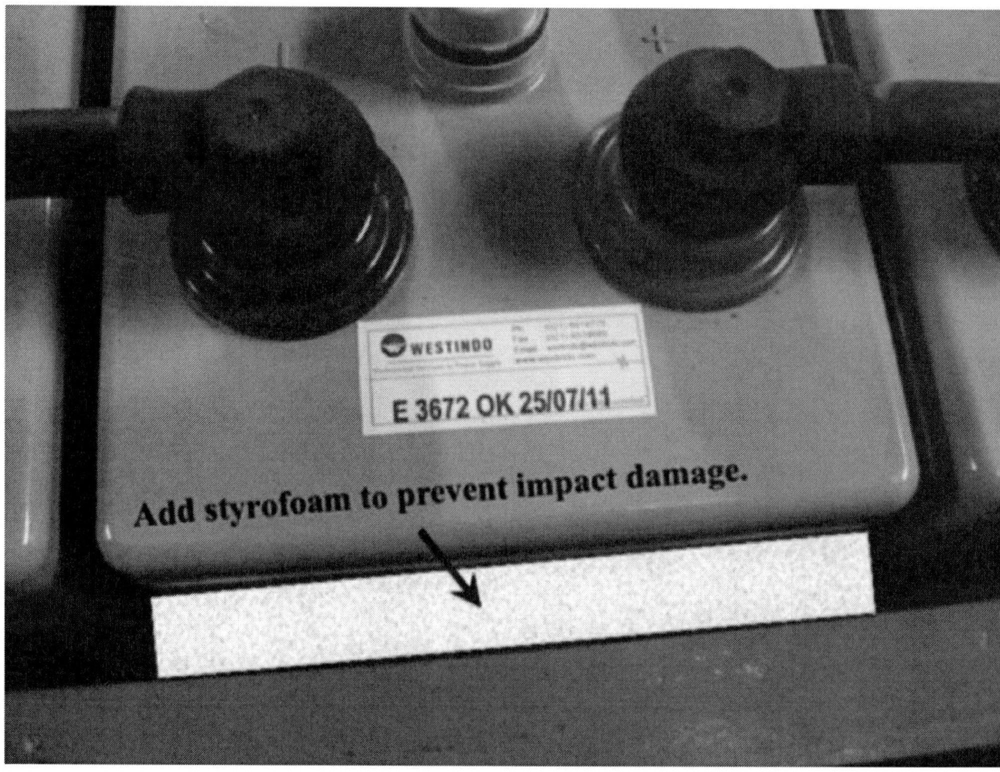

Figure 9.28 Recommended placement of Styrofoam to act as cushion to prevent impact damage during strong shaking

Figure 9.29. An indoor soccer field for employee recreation

Figure 9.30. Emergency evacuation staircase at the back of the building.

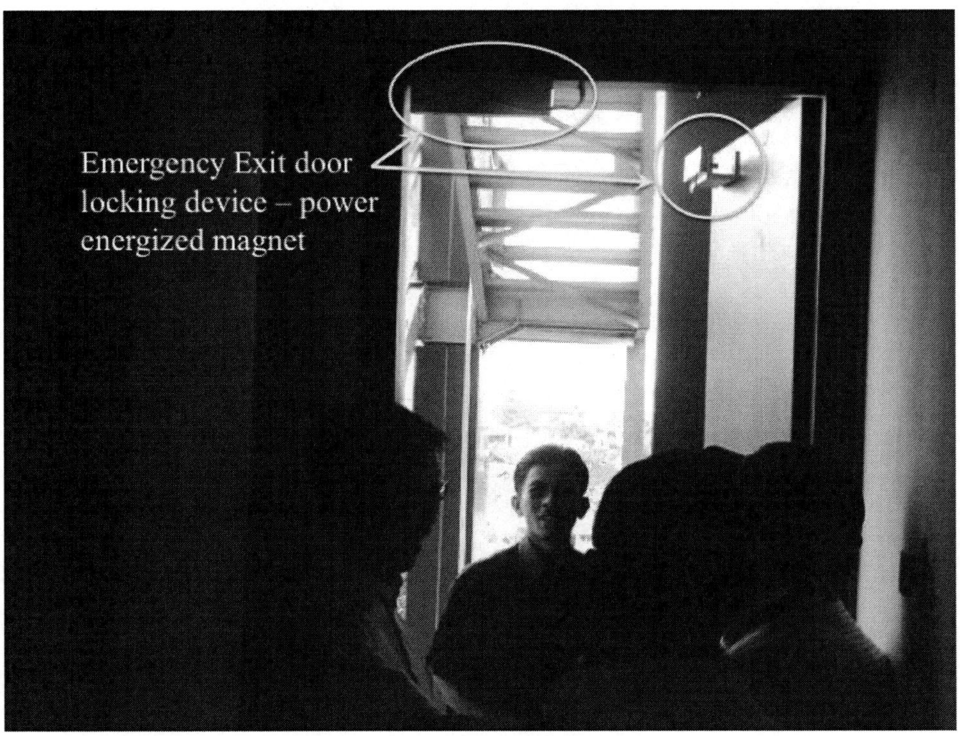

Figure 9.31. All emergency evacuation doors are locked by magnet that can only activated by power.

The team then visited the emergency power plant, which is in a separate building in the backyard of the property (Fig. 9.32). The fuel tank, which is located on the side of the building, is bolted to a steel frame (Fig. 9.33), which is anchored to a concrete pad. The anchors may not be sized correctly for the strong ground shaking of this area. Before the earthquake, there was only one back up power generator. A new backup power generator was installed to prepare for a longer commercial power outage in a post disaster situation as well as more equipment will be installed at this site (Fig. 9.34). Both units were properly installed with seismic protection in mind. However the starter batteries installation needs improvement. We felt that the two-deck battery rack should not be used (Fig. 9.35). There is enough room to place the batteries side by side on the floor, where they can be easily secured. Figure 9.36 shows the old power generator use only one set of batteries, which is installed with a steel angle frame around its base to prevent sliding during strong shaking. The day tank (Fig. 9.37) is located close to the new generator.

Figure 9.32. The power generator building with the fuel tank on the outside

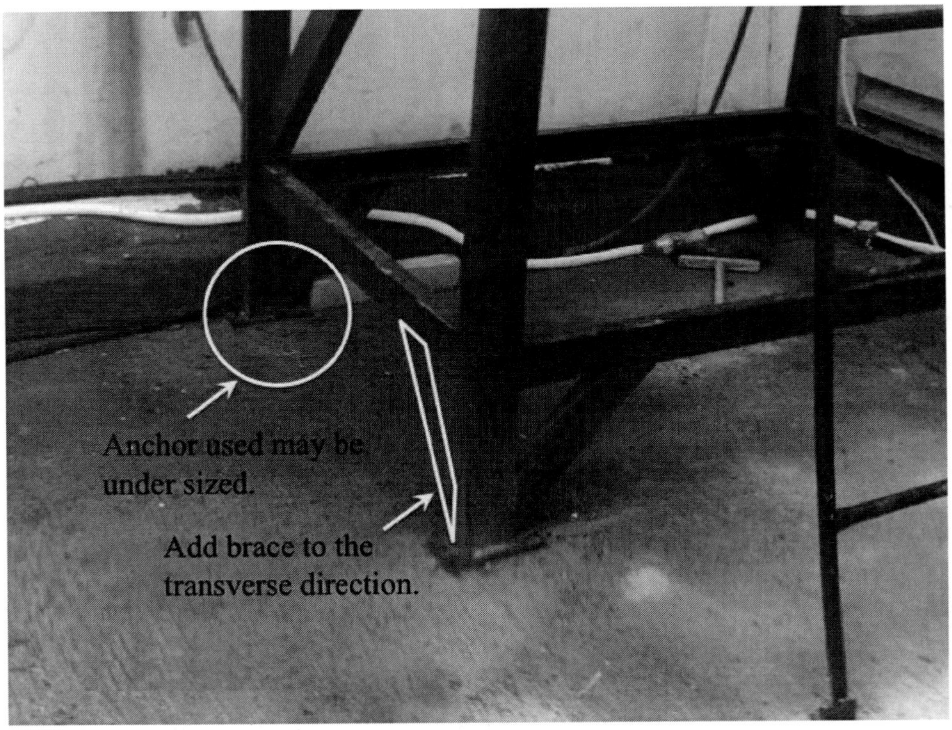

Figure 9.33. Power generators fuel tank mounting frame

Figure 9.34. The two generators with the older one (CAT) close to the wall

Figure 9.35. The starter batteries are installed on a two-tier rack, and they are not strapped to the frame. With space available, it is desirable to place the top set of batteries on the same level as the bottom set.

Figure 9.36. The older generator has one tier of batteries; however, they should be strapped to the base.

Lastly, Mr. Basyumi showed us the mobile cell sites (Fig. 9.38) parked along the front of the building. In North America these are called COWs (Cellular On Wheels), while in Indonesia they are called compact mobile base stations, or simply COMBAT. After the 2009 earthquake, 10 COMBATs were deployed. There were five available within the Padang area; the remaining five were borrowed from outside of Padang. These units were returned to their owners when the network was back to normal.

Figure 9.37. Day tank inside the power generator room

Figure 9.38. Mochamad Basyumi explaining the functions of the COMBAT to us

9.5 Schools and Shelters

With the assistance of Didi Aryadi, MSi, the assistant to the Mayor of Padang, we were able to set up two meetings with two new schools to collect information relating to school rebuilding process and the dual function of the school as a vertical shelter in case of a tsunami. The two schools that we visited were SMA Negeri 1 (a senior high school), and SMK Negeri 5 (a vocational school).

More than 200 new schools were built in the 2009 earthquake impacted area with a dual purpose–serving as both a school and a vertical shelter. They are required to withstand magnitude 9 earthquakes. All of these schools have a helipad on the roof for emergency services access. These schools are being funded by large corporations and charity groups all over the world. The structural design of these buildings is based on the funding source. The orientation of the buildings is the same; with the narrow side of the footprint facing the ocean. The intent is to reduce damage due to direct impact of tsunami wave.

SMAN1

This senior high school was funded by a Taiwan Charity Group. It is a steel structure with cast in place concrete walls. Figure 9.39 shows the front view of the school. Figure 9.40 shows the steel structure as observed in the inside of the building.

Figure 9.39. Front entrance of SMAN 1 Senior High School.

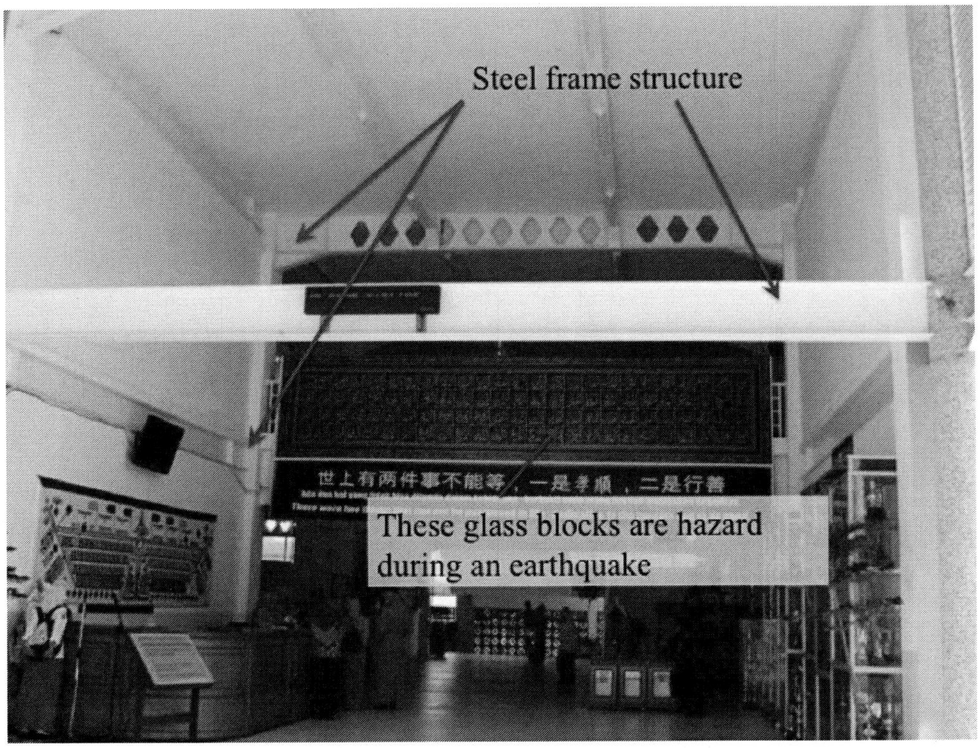

Figure 9.40. Steel beams are visible once inside the front entrance.

We were greeted by the school's head master, Dra. Wellita, who introduced a couple of his staff to us including an English teacher. She became our guide as we toured the building. There are three floors including the ground floor with a ceiling height of 8 m. There is a helipad on the roof as well as two plastic water tanks. We told the staff that flexible joints along the pipes are necessary to prevent breakage during strong ground shaking (Fig. 9.41). We noted a large speaker horn mounted on a steel truss tower (Fig. 9.42). It is used to provide advance warning of an incoming tsunami. The only drawback is that there is no access for physically challenged to go up to the roof.

After the tour, we were treated with tea and sweets. The head master presented us a gift thanking us for providing engineering advices to improve (Fig. 9.43). The gift was presented to Pat Natale Executive Director of ASCE by Alex Tang during the October 2011 TAC meeting (Fig. 9.44).

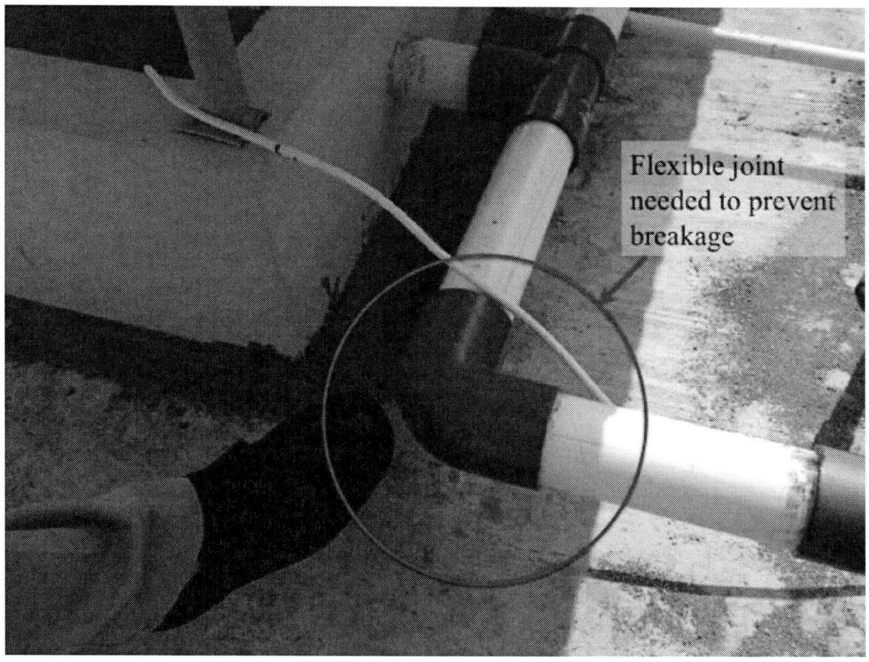

Figure 9.41. Roof top water tank piping needs flexible joint to prevent damage

Figure 9.42. Loud speaker for emergency announcements

Figure 9.43. Head Master Dra. Wellita, MM, presented to ASCE a gift received by Alex Tang.

Figure 9.44. Alex Tang presents the gift received from Head Master Dra. Wellita, MM, to Pat Natale (left) Executive Director of ASCE with ASCE President Elect 2011Kathy Caldwell (second right) and ASCE President Elect 2012 Andrew Herrmann (right). (Courtesy Frank Kim of ASCE)

SMKN5

SMKN5 is a vocational school, which occupies a large area. One of the buildings of this school collapsed during the 2009 earthquake. It was replaced by a new building funded by Chevron (Fig. 9.45). The new building holds mainly classrooms, a library, and the administration office. The building has ramp access for physically challenged (Fig. 9.46). This is a concrete frame building with infill masonry. Without a design drawing it is hard to evaluate whether it will with stand a magnitude 9 earthquake. The structure looked like a sound reinforced concrete frame structure that would be expected to withstand an earthquake. Our only concern was the infill walls, which may have shear damage during an earthquake. This could cause the infill bricks to fall and injure students. A new school was being built nearby (Fig. 9.47). This new building may be a similar structure. The head master, Drs. Yefrizon met us in his office and talked about the school's history. It is ranked as one of the best vocational schools in the Padang area. It has significant support from several industries, including car and electric appliance manufacturers. These manufacturers provided the school with their products for students to practice on.

We explored the new building with the Head Master and his assistance. Again we pointed out the need to use flexible joints for the piping connecting the water tanks on the roof. We took a group photo on the roof (Fig. 9.48).

There is the big courtyard at the back of the new building (Fig. 9.49). On the right hand side of the courtyard is a two-story building, which was renovated after the 2009 earthquake for classrooms. The left hand side houses student laboratories and workshops. When we walked through this building, we felt that it should be yellow tagged and repaired before allowing students to work inside. Figures 9.50 to 9.53 show the unrepaired structural damage from the 2009 earthquake.

Figure 9.45. The front entrance of the SMK Negeri 5 vocational school. Standing in the door way is Teuku Alvisyahrin of TDMRC and a team member of TCLEE.

Figure 9.46. The ramp for physically challenged to access the upper floors

Figure 9.47. Concrete frame building – another new school being constructed nearby

Figure 9.48. Group photo on the helipad of the SMKN 5 vocational school. From left to right, Alex Tang, Curt Edwards, Head Master Assistant, Head Master Drs Yefrizon, and Teuku Alvisyahrin.

Figure 9.49. The old part of the school with some debris still left after the 2009 earthquake

Figure 9.50. Column damaged by the 2009 earthquake not fully repaired

Figure 9.51. Another part of the workshop building was also left untouched after the 2009 earthquake.

Figure 9.52. Another part of the damaged workshop building that was not repaired.

Figure 9.53. Even the one of the column in the workshop showed concrete missing at the base.

9.6 Community

When we met with Drs. Didi Aryadi, M.Si, an assistant to the mayor of Padang, and his team (Fig. 9.54), as well as the vice-mayor (Fig. 9.55), we understood there were changes to policies and processes implemented after the 2009 earthquake. The changes are focusing on improvement and loss reduction. They appreciated that the ASCE/TCLEE returned to evaluate the changes with them. Their focus was to bring normalcy back to the community. There are many things that they really want to do, but their hands are tied by lack of funding. Although the market is almost back to normal, many buildings that were damaged by the earthquake and the fire following the earthquake were not re-constructed yet. In a country where insurance is not the norm, many owners ran out of money to rebuild, and support from the government was not nearly sufficient to fully complete the rebuilding process.

We noticed that the train was running while we were in Padang. This single railway line goes from the port to north of Padang. Both freight and passenger trains use this line.

One of the most important changes to central government funding policy is allowing unused funds to be carried over to the following year. This allows the local jurisdiction to carry out the plan without having to rush just to use up the funds.

The other change is in the building design approval process to meet building code requirements. However, an important part in this process, inspection during construction, is still missing. Without a good inspection process there is no guarantee that the construction is based on the approved design.

Another priority is rebuilding the market, which sustained extensive damage after the earthquake, including fires. Although most businesses are back in business, there are many buildings around the market that are required to be rebuilt or to be fixed before occupancy. Many of these buildings post a hazard to the people in the market. Figures 9.56 to 9.58 show the market areas with potential hazards.

The mayor's office, located at the entrance of the market, is expected to be rebuilt; Figure 9.59 shows the front of the mayor's office building.

The part of the hospital building that was severely damaged was demolished, and there is no plan to rebuild. Figure 9.60 shows the damaged part of the hospital, and Figure 9.62 shows the empty space after demolition.

Figure 9.54. Didi Aryadi and his team with TCLEE members

Figure 9.55. Vice mayor of Padang, Mr. Mahyeldi, Didi Aryadi, and TCLEE members.

Figure 9.56. Starting point to the Padang market

Figure 9.57. Fire damage to building following the earthquake

Figure 9.58. Earthquake damaged building in the market

Figure 9.59. Mayor's office building front entrance

Figure 9.60. Damaged section of the hospital building

Figure 9.61. The empty space after demolition

9.7 Buildings

Although the government provided owners with damaged properties some minimal financial support, the owners had to come with a large amount of personal funds to cover all the repairs needed or to rebuild at the same location. An additional problem was created when one neighbor repaired or rebuilt a property while the others left untouched damaged buildings that became a hazard.

Many damaged buildings were the same shape we observed in 2009. They were fenced to prevent people from entering (Fig. 9.62). However, there are many homes being constructed (Fig. 9.63), and a new bridge is being built to ease traffic congestion (Fig. 9.64). Also, there are many buildings in the process of being repaired (Fig. 9.65).

The question now is, how long it will take to recover to normalcy? Expanded financial assistance programs will be necessary to speed economic recovery to allow a faster pace of community recovery.

Through our meeting with Febrin Anas Ismail, Ph.D., of Andalas University, we understood that the building code is in the process of being updated to include a higher lateral load for building designs within Padang. The 1987 code for earthquake design was 0.07 g, then 2002 code was changed to 0.25 g after the 2009 earthquake. It is now proposed to be 0.40 g.

Dr. Febrin indicated that the problems associated with the extensive damage in 2009 were the following:

1. Substandard building materials used,
2. Missing detail design, such as hooked rebar joints between columns and beams, and
3. Poor construction techniques, lack of construction supervision, and construction procedures particularly cement bonding on columns.

He also told us that many schools and hospitals that were not severely damaged were retrofitted. During the retrofit process, they identified the problems first, such as materials used and detail designs. The university did test some base designs prior to construction.

Figure 9.62. Fenced in damaged building that is not repaired

Figure 9.63. New construction

Figure 9.64. New bridge under construction

Figure 9.65. Building being repaired

9.8 Education

We visited an NGO called Komunitas Siaga Tsunami (KOGAMI). The leader of this group was pushing to establish a curriculum on earthquake and tsunami to the primary and secondary schools in Padang to ensure that the community is educated on how to prepare for disasters and how to respond to such future events to prevent loss of lives. The school masters are very supportive and willing to organize after school classes. However, the education department is not fully convinced that it will work and is concerned about the cost.

Educating the younger generation would ensure that rare events such as large earthquakes and tsunamis can become common knowledge and will not be forgotten. We hope that KOGAMI will continue to push this good idea.

9.9 Acknowledgments

This report would not be as complete and informative without the kind support and dedicated effort of the individuals and organizations listed below. The American Society of Civil Engineers/Technical Council on Lifeline Earthquake Engineering (ASCE/TCLEE) team members are grateful for their contributions. During the course of our follow up study trip, we learned many lessons and made many friends in the lifeline arena. We all feel that sharing key information in loss reduction in disasters such as earthquake and tsunami will definitely improve losses of property and lives.

Without the financial support from ASCE this project would not have materialized. For that we thank Pat Natale, executive director of ASCE, and ASCE's Executive Committee for approving our proposal. In fact this proposal was the idea of a few members of ASCE's Executive Committee when TCLEE was requesting funding to perform the initial post earthquake investigation in Padang.

The pre-trip planning by Alex Tang, Sudi Shoja, and Teuku Alvisyahrin, Ph.D., was one of the key success factors for this project. The unparalleled support and endorsement of our effort by the local officials and the mayor's office allowed us to set up meetings with key individuals in the city to collect relevant information. Their sincere attitude and openness duly surprised us.

In addition, we really appreciate the patience and information provided by Harry Satria and his colleagues at the water treatment plant.

We are in debt and thankful to the following individuals:

Drs. H. Fauzi Bahar, Mayor of City of Padang,

Mr. H. Mahyeldi Ansharullah, SP. Dt. Marajo, Vice Mayor of City of Padang,

Drs. Didi Aryadi M.Si, Assistant to Mayor of Padang,

Mr. H. Syafril Basyir. SH, Assistant to Mayor of Padang (Board of Directors of Padang Water Department),

Drs. Dedi Henidal. MM, Chief of BPBD (Emergency response),

Mr. Ir. H. Azhar Latif, Director of Padang Water Department,

Dr. Ir. Febrin Anas Ismail, M.T., Vice Rector for Academic Affairs, Andalas University,

Mr. Mochamad Basyumi, Telkomsel Network Operations Manager, Padang,

Dra. Wellita, MM, Head Master, SMAN 1,

Drs. Yefrizon, Head Master, SMKN 5,

Dr. Teuku Alvisyahrin, Head of Professional Services Division, TDMRC Syiah Kuala University (also a TCLEE team member),

Mr. Henky Mayaguezz, BPBD (Padang Agency for Disaster Management), and

Mr. Harry Satria, Senior Engineer, Padang Water Department.

Without their willingness to share their valuable time with us, we would not have been able to achieve the goals of our trip. Again we are thankful of their support.

Unless otherwise specified all photos were taken by the authors.

Index

Page numbers followed by *f* and *t* indicate figures and tables, respectively.